CROP PHYSIOLOGY

CROP PHYSIOLOGY

Textbook for Under Graduate and Post Graduate Students

D.L. Bagdi

Department of Plant Physiology
S.K.N. College of Agriculture
Sri Karan Narendra Agriculture University
Jobner-303329
Rajasthan, India

New Delhi – 110 034

NEW INDIA PUBLISHING AGENCY
101, Vikas Surya Plaza, CU Block, LSC Market
Pitam Pura, New Delhi 110 034, India
Phone: + 91 (11) 27 34 17 17 Fax: + 91 (11) 27 34 16 16
Email: info@nipabooks.com
Web: www.nipabooks.com

Feedback at feedbacks@nipabooks.com

ISBN: 978-93-85516-30-6

Composed and Designed by NIPA

Preface

The object of this book is to bring into focus the practical application of the theoretical knowledge of crop physiology. It has been chiefly designed in accordance with the prescribed syllabus of practical part of crop physiology. It is very often seen that students familiar with the theoretical aspects of crop physiology fail to interpret the same in practice.

The precise knowledge of the crop physiological processes is of paramount importance for the undergraduate students of agriculture. This manual of Crop Physiology brought out by the Department of Plant Physiology, S.K.N. College of Agriculture, Jobner, is an appreciable step which fulfils a long felt need. This manual includes all the practical exercises of Crop Physiology (PPHYS-4221) course, offered to B.Sc. (Hons.)Agriculture students.

In this humble attempt, we strongly hope that our efforts will help users of this manual for appreciate description of basic principles and procedures given under each exercise. This manual contains 29 exercises dealing with demonstration of important physiological and metabolic processes viz. osmosis, diffusion, water potential, stomatal frequency and index, plant water status, water relation, measurement of absorption spectrum of chloroplastic pigments and fluorescence, leaf area, plant growth and transpiration. Demonstration of photosynthesis, transpiration, relationship between absorption and transpiration, conditions for seed germination, seed viability and dormancy aspects, chemical solution preparation for physiological and biochemical analysis of plant samples.

The author would like to express his gratitude to authors of those journals, manuals and books from them material has been also drawn, modified and given in this manual. The author shall be grateful to receive suggestions from readers for further improvement of this manual of crop physiology.

September, 2015
Jobner

Dr. Devi Lal Bagdi
Assistant Professor
Department of Plant Physiology
S.K.N. College of Agriculture, Jobner

Contents

Exercise

List of Colour Plates

Exercise 1

Preparation of Solutions

Object: Preparation of solutions.

Material required: Volumetric flask, measuring cylinder, beakers, pipettes, balance, distilled water and chemicals.

Theory: The use of solution is common in experimental procedures of plant physiology. It is very important for a student to be familiar with their preparation. The substance dissolved in a solution is called solute and the medium in which it is dissolved is known as solvent. Solutions of two liquids (e.g. alcohol in water) and solid in liquid (e.g. NaCl in water) are most common. Solutions are thus a homogenous mixture of a solute in a solvent.

Method of expressing the concentration of a solution: The concentration of a solution can be expressed as:-

1. Mole concentration

(i) Molar solution (M)

(ii) Molal solution (m)

(i) Molar solution: A molar solution can be prepared by dissolving any substance corresponding to one molecular weight (1 Mole) in water (solvent) so as to obtain a final volume of exactly one liter at 20° C. This will be designated as a volume molar or a molar solution (M).

The basic SI unit of quantity is the mole which gives the amount of the substance present irrespective of the volume. Mole is defined as molecular weight in grams.1 mole=molecular weight in gram=6.023×10^{23} molecules (Avogadro's number) Thus 1 mole of glucose (molecular weight 180) is 180 g.

(ii) Molal solution: Molal solution can be prepared by dissolving gram molecular weight of any substance in 1000 g of water (solvent). This will be designated as weight molar or molal solution (m). The final volume of this solution, unlike molar solution, will naturally be a little more than one liter. This increase in volume is known as the solution volume of the solute.

2. Mass concentration

(i) Percent solution (%),

(ii) Parts per million (ppm)

(i) Percent solution: A per cent solution is one which contains unit weight or volume of salt or liquid in 100 ml. of water or any other suitable solvent.

Per cent solution can be prepared normally in three ways:

(a) Weight / weight basis,

(b) Weight / volume basis,

(c) Volume / volume basis.

(iv) Parts per million solutions (ppm): When one part of a substance is dissolved in one million part of the solvent i.e. if 1 g of a substance is dissolved in 10^6 mg of water the resulting solution will be 1 ppm. In other words, mg substance per liter of solvent water is the simplest expression of ppm solution. This system is based on the fact that a liter of water weighs 1000 g or (10^6 mg) at 4°C.

$$\text{ppm solution} : \frac{W}{V} \times 10^6$$

Where W = Weight of the solute (g),

V = Desired volume (ml).

3. Solutions of acids and alkalies

(i) Normal solution (N)

Commonly used acids like hydrochloric acid and sulphuric acid are not pure and for making their solutions their actual strength has to be taken into account. Relative parameters of some acids and alkalies are given in table below;

Name of acid or alkali	Purity (% by weight)	Specific gravity at 30°C	Approximate normality
Hydrochloric acid	36.0	1.19	11.8
Sulphuric acid	96.0	1.84	36.0
Nitric acid	69.5	1.42	15.6
Glacial acetic acid	99.5	1.05	17.0
Phosphoric acid	85.0	1.71	54.0
Perchloric acid	60.0	1.54	9.2
Ammonium hydroxide	28.0	0.90	15.0
Sodiumhydroxide (saturated)	50.0	1.50	19.0

(i) Normal solution (N): A normal solution is one which contains a gram equivalent weight of a substance in one liter of solution. This can be obtained by dissolving gram equivalent weight of any substance in water so as to make the final volume one liter. This is designated as normal solution or equivalent solution.

$$\text{Equivalent weight} = \frac{\text{Molecular weight}}{\text{Total valency of cations or anions}}$$

4. Buffer solutions

Buffer resist a change in pH on the addition of acid or alkali. Certain purification procedures or reaction processes that require a particular pH are to be conducted in appropriate buffer solutions. In natural systems all the biochemical reactions of the cells are catalyzed by the enzymes whose stability as well as activity is highly dependent on the constant pH of the system. Buffer solutions have conjugate acid-base pairs. Most commonly a buffer solution consists of a mixture of a weak acid and its conjugate base e.g. a mixture of acetic acid and sodium acetate is a buffer solution. Composition of some commonly used buffers is given below;

(a) Acetate buffer (0.1M)

Stock solutions:

A.0.2 M acetic acid (11.55ml/l)

B.0.2 M sodium acetate (16.4 g of sodium acetate or 27.2 g of sodium acetate.$3H_2O$ in 1 liter)

X ml of A +Y ml of B diluted to a total volume of 100 ml

X	Y	pH
46.3	3.7	3.6
44.0	6.0	3.8
41.0	9.0	4.0
36.8	13.2	4.2
30.5	19.2	4.4
25.5	24.5	4.6
14.8	35.2	5.0
10.5	39.5	5.2
8.8	41.2	5.4
4.8	45.2	5.6

(b) Phosphate buffer (0.1M)

Stock solutions

A. 0.2 M monobasic sodium phosphate (27.8g/l)

B. 0.2 M dibasic sodium phosphate 53.65g of $Na_2HPO_4.7H_2O$ or 71.7g $Na_2HPO_4.12H_2O$ in 1 liter

X ml of A + Y ml of B diluted to a total volume of 200 ml

X	Y	pH
93.5	6.5	5.7
90.0	10.0	5.9
85.0	15.0	6.1
77.5	22.5	6.3
68.5	31.5	6.5
56.5	43.5	6.7
39.0	61.0	7.0
16.0	84.0	7.5
5.3	94.7	8.0

Method

Preparation of solutions

A. Preparation of 5% solution of NaCl and 20% solution of ethyl alcohol: Weigh 5 g of NaCl and dissolve in 95g (95ml) of distilled water. This gives 5% solution of NaCl w/w basis. If water is the solvent in which a solid is dissolved, the designation w/v is often used because 1 ml of water weighs 1 g at 4°C. A 20% aqueous solution of ethyl alcohol (v/v basis) is prepared by mixing 20 ml of ethyl alcohol with 80 ml of water. Solutions of liquids in liquids are designated as per cent solution (v/v) basis.

B. Preparation of 1 molar solution of NaCl (MW = 58.5): Weigh 58.5 g of NaCl and dissolve in a little volume of distilled water and then make the final volume exactly one litter. This gives 1M solution of NaCl. Solutions of different molarity viz. 0.01 M, 0.02, M 0.04 M and 0.08 M from 1M stock solution can be achieved by using the following formula:

$$V_1 \text{ x } C_1 = V_2 \text{ X } C_2$$

Where V_1 = The volume of the desired solution

C_1 = The concentration of the desired solution

V_2 = The volume of stock solution

C_2 = The concentration of the stock solution

C. Preparation of 1molar solution of NaCl:

Weigh 58.5g of NaCl and dissolve in 1000g (or 1000 ml) of distilled water. This will give 1 molal (m) solution of NaCl.

D. Preparation of 500 ppm solution of NaCl: Weigh 500mg of NaCl and dissolve in a little volume of distilled water and make the final volume exactly 1000ml (1L). This gives 500 ppm solution of NaCl.

Different concentrations of ppm solution viz. 10, 50, 100 and 200 ppm from 500 ppm stock solutions can be made by using the same formula mentioned for the dilution of molar solution.

E. Preparation of normal (N) solution of sulphuric acid (M.W. 98, purity 98 %, specific gravity 1.84) : Determine the equivalent weight of the acid (which is 49 for sulphuric acid).

Now calculate the original normality of the concentrated acid using the following formula:

$$\text{Normality} = \frac{1000 \text{ x specific gravity x } (\% \text{ concentration} / 100)}{\text{Equivalent weight}}$$

The computed normality of sulphuric acid above thus is 36.8 N. The final solutions of desired normality can now be prepared by simple dilution as per the following equation:

$$N_1 V_1 = N_2 V_2$$

Where

N_1 = The normality of the original acid.

V_1 = The volume of the original acid (ml)

N_2 = The normality of desired solution

V_2 = The volume of desired solution (ml)

For example, if 1 liter of 2 N sulphuric acid is to be prepared from the above original stock, we need to dilute 54.34ml concentrated acid to 1 L with distilled water (always add acid to water : do not add water to acid).The solutions prepared need to be standardized against normal sodium carbonate or borax solutions using titration method. Fill the burette with the prepared acid solution of known normality in a clean beaker. Add to it two drops of methyl red indicator. Titrate till the color becomes light pink. Record the volume of the acid from the burette.

Suppose 25 ml of borax solution (2N) is neutralized 22.5 ml of 2 N acid solutions: It indicates that there is discrepancy in volumetric measurement during the preparation of normal solution of the acid. We can easily compute the final correction with the equation $N_1 V_1 = N_2 V_2$ above since equal volumes of acid and base of similar normality should exactly neutralize each other. Thus

$$2 \text{ X } 25 = \text{N X } 22.5$$

$$N = \frac{2 \text{ X } 25}{22.5}$$

$$= 2.222 \text{ N}$$

Therefore 1 L of this prepared acid solution should be diluted to 1.11 L to obtain 2 N sulphuric acid.

Exercise 2

To Demonstrate the Phenomenon of Osmosis Using Potato Osmoscope

Object: To demonstrate the phenomenon of osmosis using potato osmoscope.

Material required: A potato tuber, sugar solution, beaker, water, petridish, knife, measuring cylinder, and stand.

Theory:Osmosis is a phenomenon in which water molecules diffuse through the semi permeable membrane from region of its higher concentration to the region of lower concentration.

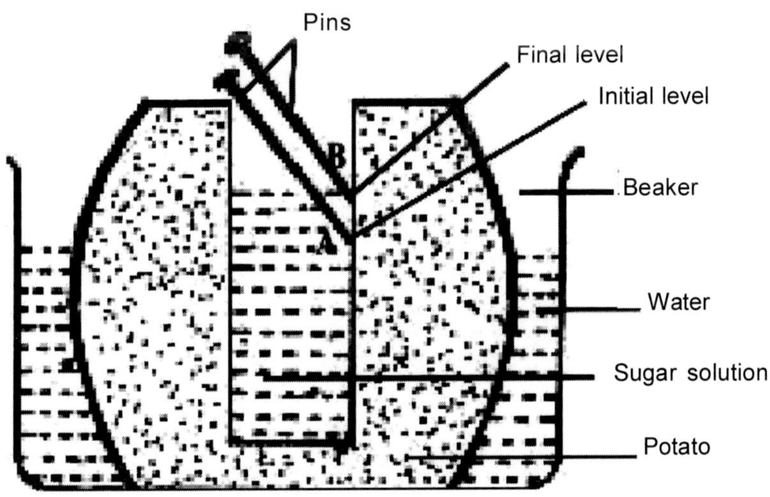

Fig. 1: Potato osmoscope

Method: A potato tuber is taken and its skin is peeled off. Then a cavity is made with the help of a knife. A petridish is taken and filled up with water with a little amount of safranin (to make the water red).The cavity of tuber is filled with strong sugar solution and it is kept in the petridish (containing colored water). The whole experiment is run for few hours and observe.

Observations: The initial level of sugar solution in the cavity of tuber rises after some times.

Results: Water entered in the cavity because of osmosis through the tuber. Thus tuber acts as a semi permeable membrane. The entry of water from petridish to potato cavity shows endosmosis because the level of water into potato cavity increased.

Precaution

1. There should be difference in level of water and sugar solution.
2. The skin of potato tuber is properly removed.
3. Potato tuber should be fresh.

Exercise 3

To Demonstrate the Phenomenon of Endosmosis and Exosmosis by Using Grapes and Raisins

Object: To demonstrate the phenomenon of endosmosis and exosmosis by using grapes and raisins.

Material required: Grapes, raisins, beaker or petridish, water and sugar solution.

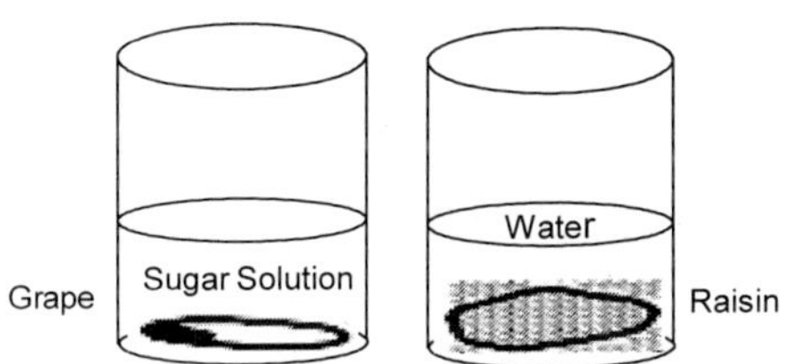

Fig. 2. Phenomenon of endosmosis and exosmosis

Theory: If a cell is placed in a solution containing more solute concentration than cell sap, water diffuses out of the cell and the cytoplasm shrinks away from the cell wall. This process is called exosmosis. On the contrary if a cell is placed in a solution containing less solute concentration than cell sap, the water diffuses into the cell and its volume increases. This phenomenon is called as endosmosis.

Method: Take some raisins and put them in water for 5-6 hrs. Then observe the change in size and appearance of raisins. Similarly in a saturated solution of sucrose, put some grapes in it and observe.

Observations: The raisins placed in water swell up and increase in volume. And the grapes which are in sugar solution shrink and decrease in size and volume.

Results: The increase in size and volume of raisins is due to endosmosis, because the solute content of raisins was of higher concentration than that of water i.e. flow of water is toward raisins. Similarly, the shrinkage or decrease in size and volume of grapes is due to loss of water i.e. by exosmosis. Since they are placed in concentrated solution of sugar, meaning, the flow of water is from grapes to sugar solution.

Precautions

1. The raisins should be dry
2. Petridish should be filled with water and sugar solution separately.
3. Grapes should be fresh and healthy.

Exercise 4

To Demonstrate the Process of Diffusion

Object: To demonstrate the process of diffusion.

Material required: Beakers, copper sulphate, distilled water and sugar solution.

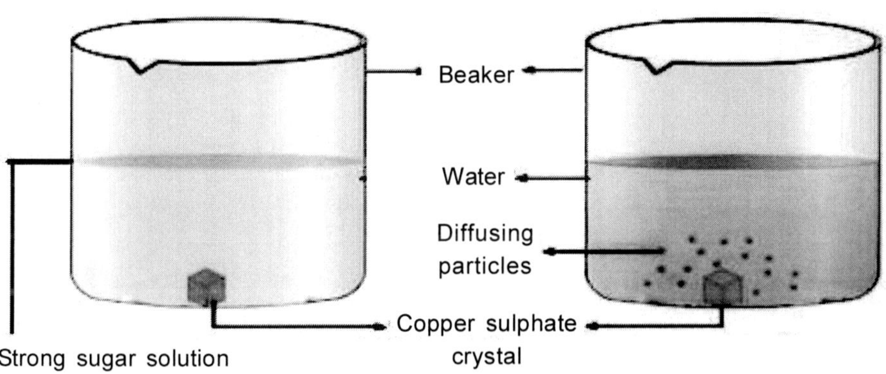

Fig. 3. The process of diffusion

Principle:The tendency of molecules of gases, liquids and solids to get evenly distributed throughout the available space is called diffusion. It is due to the fact that molecules or ions or colloidal particles are in a state of perpetual motion due to the kinetic energy present in them. They thus colloid with one another and ultimately get deflected with the direction of least correction.

Procedure

1. Take a beaker and fill it about two third with strong sugar solution. Now place a crystal of copper sulphate in this beaker and wait for some time.

2. In another beaker pour water about two third and place a crystal of copper sulphate in this beaker and wait for some time.

Observations

In second demonstration after some time, the crystal with no longer be visible but the whole of the water in the beaker will turns to blue color. In first demonstration, the crystal of copper sulphate is visible for a longer time in strong sugar concentration and the color of strong sugar solution is only lightly turns to blue color.

Result

Diffusion or observable movement of the particles from one place to another can takes place only when there is difference in the concentration of a substance in the different parts of a system.

Precautions

1. The beaker should not fill more than two third portions.
2. The sugar solution must be strong.

Exercise 5

Methods of Measurement of Water Status and Water Saturation Deficit in Plant Parts

Object: Methods of measurement of water status and water saturation deficit in plant parts.

Material required: Leaf samples, leaf borer, water, petridishes, forceps, oven and blotting paper.

Theory: Plant water status strongly influences plant growth and biomass production through its effects on leaf and root expansion and on photosynthesis. Therefore, measurement of plant water status is an important component of understanding plant processes and maximizing yield under different ecosystems.

Method: Plant water status can be measured and computed by any one of the following methods

Fresh weight method: Prepare a humid chamber to bring the leaf samples from the field to laboratory. Blot the leaf surfaces for removing the excess water. Then with the help of cork borer or leaf borer cut the leaf discs (about 25) and immediately note fresh weight. Then keep the weighed leaf discs into an oven by placing in a petridish at 70^0 C for 24 hours and record its dry weight until the weight remains constant.

Observations: The water content of the tissue is expressed on fresh weight basis by using the formula:-

$$\% \text{ Water Content} = \frac{\text{Fresh weight - Dry weight}}{\text{Fresh weight}} \times 100$$

(b) Dry weight basis: Take the fresh weight and dry weight of the plant material as above and express the water content on dry weight basis using the following formula:

$$\% \text{ Water Content} = \frac{\text{Fresh weight - Dry weight}}{\text{Dry weight}} \times 100$$

(C)Relative water content (R.W.C.) and water saturation deficit (W.S.D.):

Method: Select normal and fresh leaves. Cut the leaf discs with the help of leaf borer and immediately take fresh weight. Then float leaf discs on water contained in petridishes, water uptake and allow them for attaining maximum turgidity. After an optimal period (4 -6 hours) take out the discs on filter paper and blot them gently so that water on disc surface is removed

and take the turgid weight. Then put the leaf discs in a paper envelope (made of filter paper) and oven dry at 70°C till constant weight is obtained. Record the observations and calculate R.W.C. and W.S.D. using the following formulae:

$$\% \text{ R.W.C.} = \frac{\text{Fresh weight - Dry weight}}{\text{Turgid weight - Dry weight}} \text{ X } 100$$

$$\% \text{ W.S.D.} = \frac{\text{Turgid weight -Fresh weight}}{\text{Turgid weight – Dry weight}} \times 100$$

Observation Table: For R.W.C. and W.S.D.

S.No.	Plant part	Fresh wt.(g)	Turgid wt.(g)	Dry wt.(g)	R.W.C.(%)	W.S.D.(%)
1	Leaf					
	Root					
	Stem					
2	Leaf					
	Root					
	Stem					

Results: The R.W.C. and W.S.D. of the given leaf sample is --------- % and --------- % (for the above sample observation table.

Precautions

1. Leaves must be fresh and partially mature.
2. While cutting the discs, midrib should be removed.
3. Only interveinal area should be taken.
4. Delay in taking fresh weight of leaf discs should be avoided to prevent moisture loss from discs by way of transpiration.

Exercise 6

Estimation of Water Potential of Plant Tissue by Chardakov's Method

Object: Estimation of water potential of plant tissue by Chardakov's method.

Material required: Test tubes, sucrose or polyethylene glycol (PEG) 6000, methylene blue and dropper.

Principle: Concentration of the solution in which a plant tissue is immersed changes. If the solution is hypertonic (with lesser ψ_w) the tissue loses water and the solution becomes dilute. If the solution is hypotonic (with higher ψ_W) the tissue absorbs water and the Solution becomes concentrated. If the solution is isotonic (with ψ_w equal to the tissue sap) the tissue neither loses nor absorbs water and the solution concentration does not change. Thus the ψ_w of this solution gives the ψ_w of the tissue.

Procedure and observations

Prepare a graded concentration series of sucrose or PEG 6000 of different molarities i.e. 0.05, 0.1, 0.15, 0.2, 0.25, 0.3. Take these solutions in two set of test tubes each. Color one set of each solution by dissolving a small crystal of methylene blue dye in it (control solution).This does not change the ψ_w of the control solution significantly with respect to its non-colored counterpart (test solution).Immerse tissue samples (approximately of equal size) in each colorless test solution test tube. Allow the tissue to come to equilibrium with the solution for about 15-20 minutes at room temperature. Remove the tissue from the test tubes. Gently release a drop of the corresponding colored control solution in the colorless test solution. If the control solution drop rises up in the test solution in

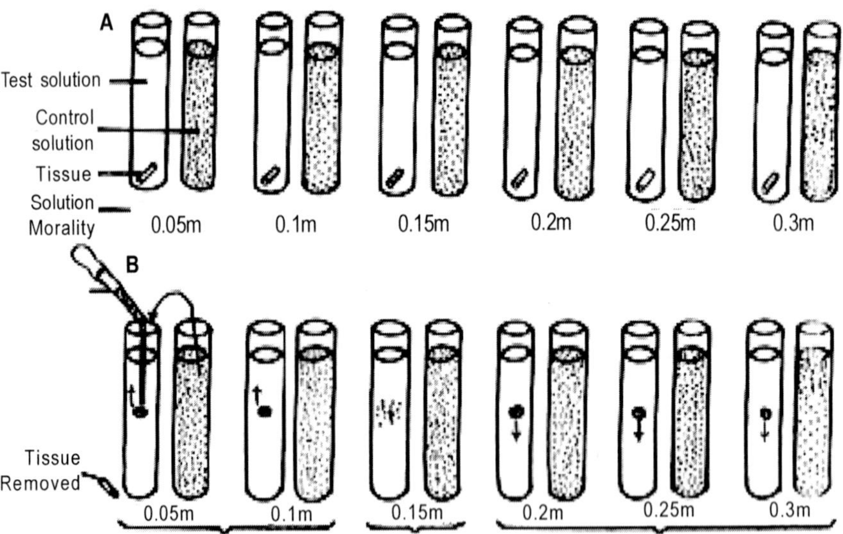

Fig.4: Measurement of water potential of plant tissues by Chardakov's method

which the tissue was incubated the test solution has become denser, indicating that the tissue has taken up water. If the drop sinks,

A: The tissue placed in test solution

B: Drops of control solution released in test solution the test solution has become less dense, having absorbed water from the tissue. If the drop diffuses evenly in to the test solution without rising or sinking, then no change in the concentration has occurred and the water potential of the solution equals that of the tissue. Calculate ψ_w of this solution using Van't Hoff equation. Chardakov's method is very simple and fairly sensitive for determination of the ψ_w of plant tissues even in the field. Record the observations in note book and calculate the tissue water potential as described above.

Data table

S.No.	Molarities of solution	Direction of movement of drop*
1.	0.05	
2.	0.1	
3.	0.15	
4.	0.2	
5.	0.25	
6.	0.3	
7.	0.35	

*rises up; diffuses; sinks

Exercise 7

Estimation of Photosynthetic Pigment from Plant Leaves

Objective: Estimation of photosynthetic pigment from plant leaves.

Material required: Plant leaves, test tube, test tube stand, spectrophotometer, balance, centrifuge machine, mortar and pestle, acetone and distilled water.

Theory: Photosynthetic pigments, namely, the chlorophylls are present in green leaves and shoots of plants. There are at least five types of chlorophylls in plants. Chl'a' and 'b' occur in higher plants, ferns and mosses. Chl 'c','d' and 'e' are found only in algae and in certain bacteria. Chlorophylls are extracted from plant leaves. The absorption peaks of chl 'a' at 663 nm and 'b' at 645 nm are characteristic measures from which amount of chlorophylls can be calculated.

Method: Weigh 100 mg leaf samples and homogenized in 10 ml, 80% acetone with the help of mortar and pestle. The homogenate transfer to the measuring cylinder and volume is made to 10 ml with 80% acetone. This homogenate is centrifuged at 5000rpm for 10 minute. The absorbance of the supernatant is measured at 663 and 645nm in spectrophotometer.

Observations: Optical densities at $\lambda = 663$ nm and $\lambda = 645$ nm are recorded as follows:

S.No.	OD at 663 nm	OD at 645 nm
1		
2		

Calculations: Calculate amount of chlorophylls present in the 10 ml extract as mg chl/gm tissue as follows:

(1) mgs chl 'a' /g = 12.7 (A663) – 2.69 (A645) x V/ 1000 x W

(2) mgs chl 'b'/g = 22.9 (A645) – 4.68 (A663) x V/1000 x W

(3) total chl /g tissue = 20.2 (A645) + 8.02 (A663) x V/1000 x W

Where

A = Absorbance at specified wavelengths

V = Final volume of chl extract in 80% acetone

W = Fresh weight of tissue extracted

Result:The total chlorophyll content obtained from leaf samples grinding in 80% acetone solvent is= mg/g fresh weight of leaf.

Content of chlorophyll 'a' = mg/g fresh weight of leaf.

Content of chlorophyll 'b' = mg/g fresh weight of leaf.

Precautions

1. Use tissues that are only light green in color.
2. Handle acetone solvent carefully.
3. Put the spectrophotometer on 15 minutes before taking observations.
4. Set one wavelength at a time and use 80% acetone solvent as the instrument blank.
5. Understand operations of spectrophotometer before switching on and off the different modes available.

Exercise 8

Measurement of Absorption Spectrum of Chloroplastic Pigments

Object: Measurement of absorption spectrum of chloroplastic pigments.

Material required: Spectrophotometer, plant leaves, ethanol, electric balance, beaker, test tubes and water.

The absorption spectrum

Since solutions of pure substances do not absorb the energy of all wavelengths of light equally, a substance may be identified by the unique pattern of wavelengths absorbed. The chlorophylls in plants absorb strongly in the blue wavelengths (about 450nm) and red wavelengths (about 650nm) but reflect the green wavelengths (about 525nm). A plot of absorbance versus visible wave lengths (400 to 700nm) for a solution of chlorophyll a shows two major peaks, one at 450 and one at 650 nm, and a valley from 500 to 625nm (See Figure below) This spectrum is characteristic for chlorophyll a and may be used as an aid in its identification.

By measuring the absorbance of an uncharacterized solution over a range of wavelengths and plotting the absorbance value on the Y-axis and the wavelength on the X-axis, one can determine the absorption spectrum of a sample. The absorption maximum of any pure substance in solution is the wavelength where absorption is the greatest.

Method: Extracting Chlorophyll from Spinach

1. Shred a medium-sized leaf of spinach and mix with 10 ml of ethanol in a 50 ml beaker.

After approximately 5 minutes, the solution should have a green tint.

Note: If the Solution looks yellow, the extraction is not finished.

2. Filter the mixture of leaves and ethanol into another 50-ml beaker, using a funnel with a fluted filter. Discard the residues;Spectrum of Chlorophylls.
3. Fill one cuvette with ethanol (the blank) and another with the solution of chlorophylls (the sample). First, check the absorbance at 440nm. The reading there, where both Chlorophylls absorb to some extent, should be somewhere between 0.3 and 0.8. If the solution is too concentrated, pour the sample into a graduated cylinder and add an equal amount of alcohol to dilute the solution by half. Check the absorbance at 440nm again. Dilute further if needed. record the dilution needed.
4. Set the wavelength dial to 400 nm and measure the % transmittance of your sample. Continue to measure the % T every 10nm between 400nm and 700nm, remembering that adjust the instrument to full scale at each new wavelength. You can take readings every 5nm between 420 and 480nm to make it easier to observe the absorption due to both chlorophylls. Convert % T to A. Plot absorbance vs. wavelength by drawing a smooth line through the points (don't "connect the dots").

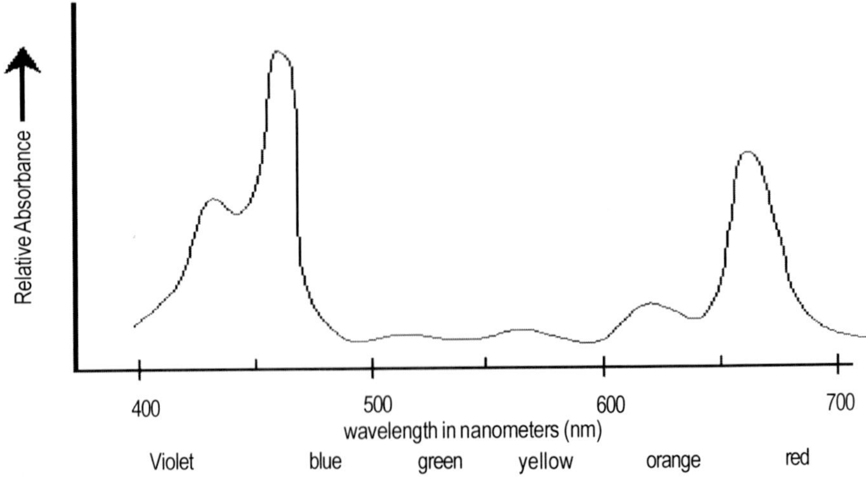

Fig. 5. Absorption spectrum of chloroplastic pigments

Observation

Spectrum of Chlorophylls

Dilution (if required): Vol. Leaf Extract ___ ml Vol. Alcohol added: ____ ml

λ nm	%T	A	λ nm	%T	A
400			530		
410			540		
420			550		
425			560		
430			570		
435			580		
440			590		
445			600		
450			610		
455			620		
460			630		
465			640		
470			650		
475			660		
480			670		
490			680		
500			690		
510			700		
520					

%T=Percent transmittance A=Absorbance λ=Wavelength

Results

Compare the spectrum with the one above to find evidence for presence of both chlorophyll **a** and **b**.

Estimate the relative amounts of a and b in your sample.

Precautions

1. Switch the spectrophotometer for 30 minute before taking reading of absorbance and %T.
2. Cuvette of spectrophotometer should be neat and clean.
3. Handle reagent carefully.

Exercise 9

To Demonstrate Chlorophyll Fluorescence

Object: To demonstrate chlorophyll fluorescence.

Material required: Spectrophotometer, fresh green spinach, mortar, pestle, acid washed sand, acetone, funnel, filter paper and test tubes.

Principle: Chlorophyll molecule absorbs light in the blue (450nm) and red (660nm) region and get into excited state and emits an electron. The electron drops back to the ground state within 10^{-9} seconds and some energy (which is not used in biochemical reaction) are given out as the observed fluorescent far red light.

Procedure and observation: Take some soft green spinach leaves. Grind these with a pestle -mortar in acetone adding a pinch of acid washed sand. Make a fine paste. Filter to get a clear solution in a test tube. Take the test tube in bright sun light.

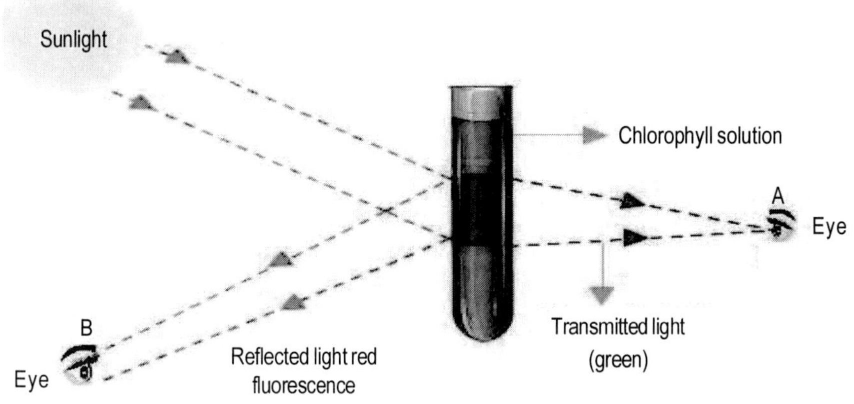

Fig. 6. Demonstration of chlorophyll fluorescence

Observed the transmitted light through the test tube i.e. with the solution between the source of light and the .The solution appears green. Now observe the reflected light, i.e. with the source of light behind the eye.The solution appears red. This phenomenon is due to fluorescence.

Precautions

1. Leaves should be fresh and soft green.
2. Acetone should be handling carefully.
3. Test tube should be properly washed or new.

Exercise 10

Measurement of Leaf Area by Graph Paper Method

Object: Measurement of leaf area by graph paper method.

Material required: leaves, graph paper and pencil or pen.

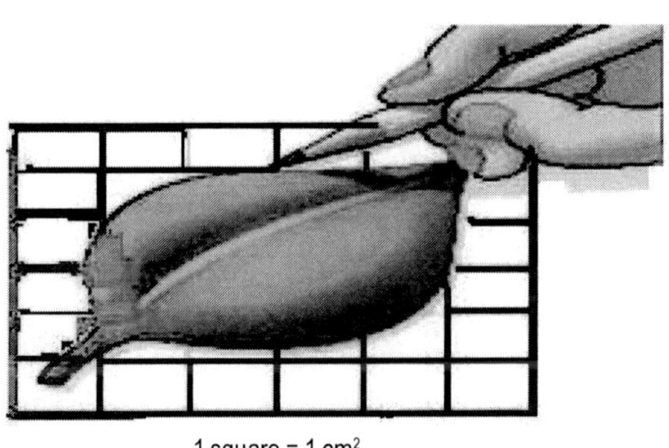

1 square = 1 cm^2

Fig. 7. Graph paper

Theory: Leaf area is an important measurement in bio productivity studies, as it bears important and direct relationship to photosynthesis and transpiration. Leaf area is also important in determining the percentage of solar radiation intercepted by an individual plant or crop canopy. Photosynthesis centers on individual leaves. However productivity stand of vegetation depends not only on individual leaves but on canopy structure. The photosynthetic productivity of a canopy depends on both the amount of leaf present and the configuration of the leaves making up the canopy. This leaf area directly influences plant growth and final yield.

Method: Keep and press the unfolded leaves on graph paper and count the large and small squares covered by the leaves to estimate the surface area for each leaf. Repeat this for each leaf of a plant randomly treatment wise of any experiment.

Results: The estimated leaf area of the given leaves is .... cm^2.

Precautions

1. Leaves should be fresh.
2. There should be no folding in the leaves.
3. The counting of large and small squares of the graph paper should be exactly.

Exercise 11

Estimation of Stomatal Index and Stomatal Frequency with Replica Method

Object: Estimation of stomatal index and stomatal frequency with replica method.

Material required: Leaf sample, blue/red correction fluid, transparent finger nail polish, brush glass slide and microscope.

Principle: The number of stomata per unit area of the leaf is called as Stomatal Frequency. The stomatal index is defined as the percentage number of Stomata as compared to all the epidermal cells in a unit area of leaf i.e.

$I=(S/E+S) X 100$

Here I, S and E stand for stomatal index, number of stomata per unit area and the number of epidermal cells per unit area, respectively. The total pore area when stomata are fully opened is 1-2% of the total leaf surface.

Procedure: The stomatal index and frequency can be estimated by replica method as suggested by Wolf and his associates (1979).Blue/red correction fluid or transparent finger nail polish (collodion) is used as replicating material. Fluid dipped brush is stroked across the area to be replicated with a single motion. After few seconds when fluid become firm, the replica can be removed. A small piece of celluloid tape is secured to a glass slide. The exposed sticky surface is then placed over the replica and with slight pressure; the film is removed from the leaf surface. The replica is mounted for viewing as imprint on reverse side of the image of the leaf surface. Using Microscope with pre-calibrated grid, the number of stomata and epidermal cells can be counted.

Data table: Stomatal number, epidermal cells and stomatal index of some field crops.

Crop	Stomatal Number	Epidermal cells	Stomatal index (%)
Bean	4	12	250
Arhar	5	12	29.4
Cotton	6	32	15.8
Groundnut	9	36	20.0
Sorghum	8	20	28.6
Tomato	7	25	21.9

Result: Different plant species differ in stomatal number and stomatal index.

Precautions

1. The leaf should be fully mature.
2. The lens of microscope should be neat and clean.
3. The place where microscope is placed, sunlight should be there.
4. Reagents should be handled carefully.

Exercise 12

Demonstration of Leaf Anatomy or Structure of C_3 and C_4 Plants (Demonstration by Already Prepared Slides)

Object: Demonstration of leaf anatomy or structure of C_3 and C_4 Plants (demonstration by already prepared slides).

Material required: Already prepared slides of corn and wheat leaves and microscope.

Demonstration of C_3 and C_4 Leaf Anatomy

By looking at their anatomy, in C_3 plants, bundle sheath cells do not contain chloroplasts; carbon fixation and Calvin Cycle reactions occur in mesophyll cells (and in the presence of oxygen). In C_4 plants, the bundle sheath cells contain chloroplasts; carbon is fixed in mesophyll cells and then transported to bundle sheath cells where Calvin Cycle reactions occur in the absence of oxygen. In both, photosynthesized sugars then enter the plant's vascular system.

C_4 have a concentric arrangement of the bundle sheath and mesophyll layer, the bundle sheath is also thicker. Another difference is their interveinal distances, from one bundle sheath to another you have in C_4 only around 4 mesophyll cells but in 3 they are separated by 12.

Overall, C_4 plants are more adapted to environments with more oxygen, and C_3 plants are more adapted to environments with more carbon dioxide.

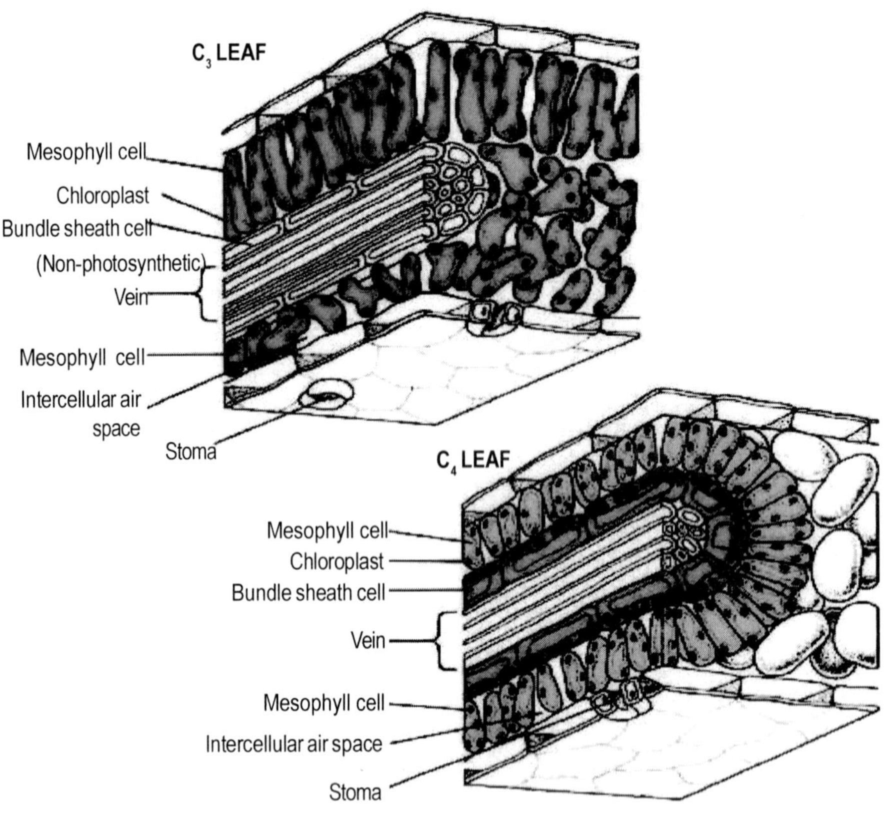

Fig. 8. Leaf structure of C_3 and C_4 plants

Precautions

1. Microscope should be properly operated.
2. Glass slide and cover slip should be neat and clean.
3. Staining solution should be fresh.
4. Microscope should be place near window, where light should be available.

Exercise 13

To Demonstrate that Water is Lost from a Living Plant During Transpiration

Object: To demonstrate that water is lost from a living plant during transpiration.

Material Required: A potted healthy plant, a bell jar, a rubber sheet or oil cloth, a glass sheet and grease.

Theory: Transpiration: Loss of water in the form of water vapour from the internal tissues of living plants through aerial parts of the plant is known as transpiration.

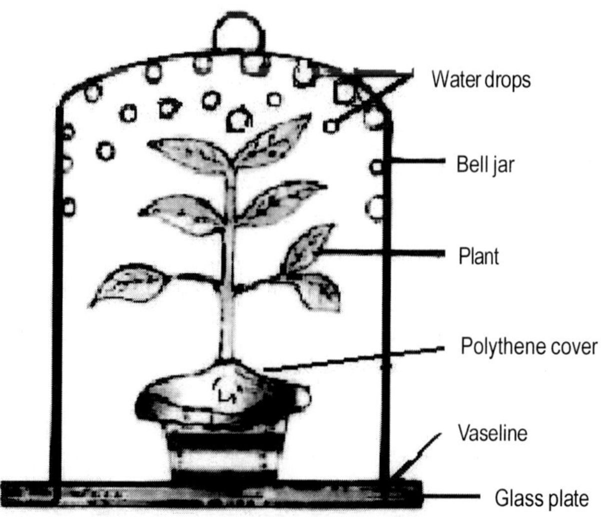

Fig. 9. Demostration of water loss by bell-jar

Method: A potted plant is taken and the soil surface of the pot is properly covered by an oil paper to check the direct evaporation from the soil surface. The potted plant is covered by a previously dried bell jar and is left in this condition for about one hour. Observations are made after sometime.

Observations: After some time it is seen that, in the inner wall of the bell jar some drops of moisture/water are deposited.

Results: The Water drops that appear on the inner wall of the bell jar come from the plant by phenomenon of transpiration.

Precautions

1. By putting grease at the rim-base of the bell jar it should be made air tight.
2. Before starting the experiment the plant should be well watered.
3. The pot and the exposed soil should be covered with polythene.
4. Bell jar should be made of glass.

Exercise 14

To Measure Rate of Transpiration of Potted Plant by Gravimetric (Lysimeter) Method

Object: To measure rate of transpiration of potted plant by gravimetric (Lysimeter) method.

Material required: Potted plants, Polythene sheet, Platform and electric balance etc.

Method: Select same size and same type well watered potted intact plants. Seal the pots and the root system with a sheet of polythene so that the water loss to the atmosphere occurs only from the shoot system not from the soil (To check evaporation losses). Immediately record the weight of the potted plant and keep them in different environment such as (1) warm environment (2) cool environment (3) Calm environment.

Now record the weight of the entire pot periodically at one hour interval over specified period (final weight) for each environment.The loss in weight represents the quantity of water transpired. Express the rate of transpiration grams water vapor loss per plant per hour.

Observation table

S.No.	Treatment	Initial weight of the potted plant	Weight at hourly interval			Average water loss(g)
			Ist hour	2nd hour	3rd hour	
1	Warm environment					
2	Cool environment					
3	Calm environment					

Calculation: Determine the amount of water loss by subtracting the weight from the initial weight and calculate the amount of average water loss per plant at the end of the experiment. This will calculate the water vapor loss per plant per hour.

Result: The rate of transpiration (gH_2O $hour^{-1}$ $plant^{-1}$)

1. Warm environment ....

2. Cool environment ....

3. Calm environment ....

Exercise 15

To measure the Rate of Transpiration by Using Ganong's Potometer

Object: To measure the rate of transpiration by using Ganong's potometer

Material required: Ganong's potometer, beaker, wax or vaseline, stop watch and twig or a branch of a plant.

Theory: The amount of water absorbed is almost equal to the amount of water transpired. The difference between the two amounts is negligible.

Method: Potometer consists of wide mouthed tube fitted with a hard cork in which a twig or branch of a plant is inserted. Other end of tube is dipped in beaker. One half of the tube is graduated. The apparatus is filled with water using the reservoir. An air bubble is inserted in the tube as shown in figure. Duc

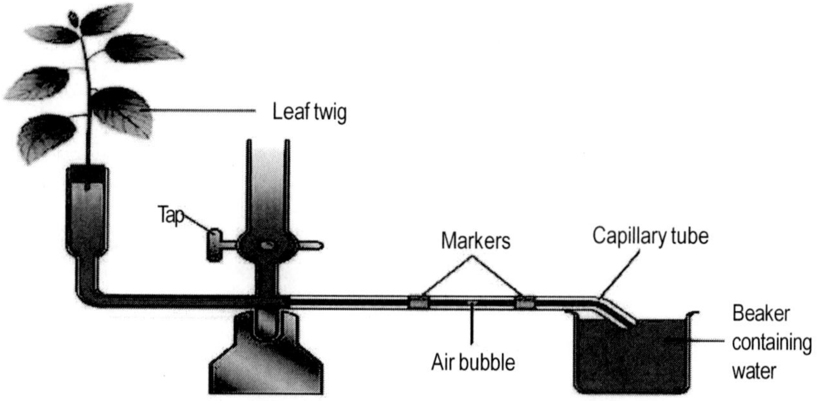

Fig.10: Ganong's potometer

to transpiration water is absorbed by twig. Thus, the air bubble moves towards the twig. The movement of air bubble in the horizontal tube is noted and distance traveled is calculated at definite intervals of time and rate of transpiration is determined in terms of ml/hour/area. The volume of water can be calculated by the formula:

$$V = \pi r^2 I$$

Where V = Volume, π = 22/7, r = radius of graduated tube and I = length or distance traveled by bubble.

Observations: Observations are made repeatedly (by backward sliding of bubble by adding water through reservoir) under wide range of conditions.

Result: The bubble moves in horizontal tube and the difference in reading give the rate of transpiration. The absorption of water measured is proportional or the amount of water lost in transpiration. Hence the rate of ration is measured.

Precautions

1. The shoot must be cut under water to avoid blocking of vessels by air.
2. Apparatus should be air tight.
3. The twigs must have more leaves.

Exercise 16

To Find Out the Relation Between Transpiration and Absorption

Object: To find out the relation between transpiration and absorption.

Material Required: Wide mouthed glass bottle, with a narrow graduated side tube, split rubber cork, water and a plant with intact root system.

Theory: The loss in weight of plant indicates the amount of water transpired. And fall in water level of the side tube indicates the amount of water absorbed. This shows that amount of water transpired is normally equal to the amount of water absorbed by the plant roots.

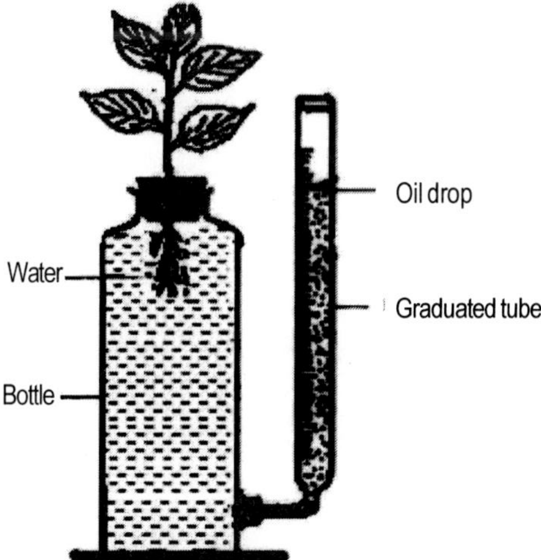

Fig. 11. Comparison of transpiration and absorption of water

Method: Take a wide mouthed glass bottle with an opening at the side to which graduated tube is fixed. Fill the apparatus completely with water and put few drops of any oil on the water level of the graduated tube to avoid water loss due to evaporation. Fix a young vigorously growing plant with its root system immersed in water by means of spliced rubber cork and make it air tight as shown in figure. Weigh the apparatus in a open pan balance and note the weight. Record the initial reading of the water level in the side tube and the experiment is run for an hour and observed.

Observations: After one hour reweigh the whole apparatus and record the final level or reading of the water level in the side tube.

Results: As the water is absorbed by the roots, the water level in graduated tube decreases. This is the amount of water absorbed by the roots. The difference in first and second weight gives the amount of water transpired which is normally equal to amount of water absorbed.

Precautions

1. The whole apparatus should be air tight.
2. Take a vigorously growing plant with intact root system.
3. Oil drops should be added to prevent water loss due to evaporation in a side tube.

Exercise 17

To Demonstrate Unequal Transpiration by the Two Leaf Surfaces from Four Leaf Method

Object: To demonstrate unequal transpiration by the two leaf surfaces from four leaf method.

Material required: four leaves of china rose or any dorsiventral leaves, two stands, thread and grease or vaseline.

Theory: Transpiration is a process in which loss of water occurs in the form of water vapour from the aerial parts of the living plant.

Method: Take four leaves of equal size from a plant (e.g. China rose). Now apply or smear the vaseline on the four leaves as follows:

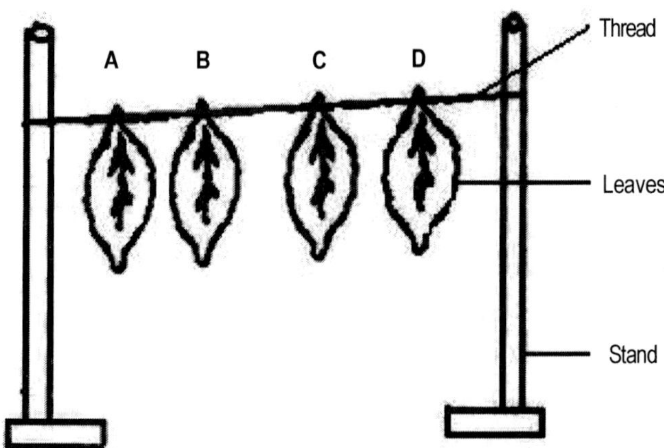

Fig. 12. Comparison of transpiration rates

(i)	On Leaf 'A'	=	Don't apply any Vaseline or grease.
(ii)	On leaf 'B'	=	Apply vaseline on upper surface.
(iii)	On Leaf 'C'	=	Apply vaseline on lower surface.
(iv)	On Leaf 'D'	=	Apply vaseline on both the surfaces.

Weigh them carefully and hang them side by side as shown in figure. After a day, weight them again and note the differences in weight and appearance of the leaves.

Observations: The observations made are as follows:

Leaf 'A': It has wilted completely and shows considerable loss in weight.

Leaf 'B': It has nearly wilted and its weight has decreased.

Leaf 'C': It has no wilted i.e. fresher than A and B leaves.

Leaf 'D': It is freshiest of all four leaves.

Results: Since leaf A does not have any Vaseline, transpiration takes place rapidly from the two surfaces, through stomata so it completely dries or wilts. On leaf B, Vaseline was applied on upper surface only so it dries faster. On leaf C, Vaseline was applied on lower surface so it dries slowly than leaf B. On leaf D, Vaseline was applied on both surfaces, so no transpiration occurs and the leaf remains fresh for longer period. It proves that the rate of transpiration through stomata is faster than cuticular transpiration.

Precautions

1. The leaves should be of same size and of same plant.
2. The leaves should be properly smeared with vaseline.

Exercise 18

To Demonstrate that Light and Carbondioxide are Essential for Photosynthesis (Moll's Half Leaf Experiment)

Object: To demonstrate that light and carbon dioxide are essential for photosynthesis (Moll's half leaf experiment).

Material Required: A potted plant, a wide mouthed bottle, caustic potash, splits of rubber Cork and vaseline etc.

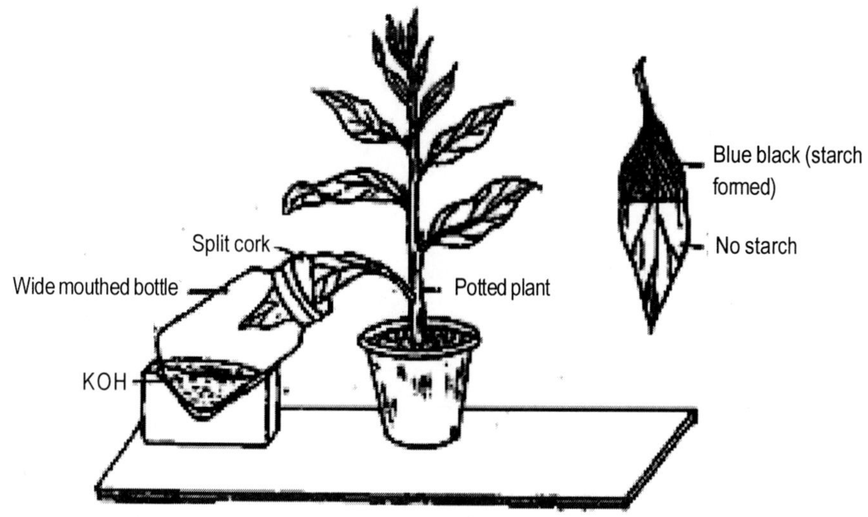

Fig. 13. Moll's half leaf experiment

Theory: Photosynthesis is the process by which CO_2 of the air is converted into the organic matter of the green plants with the aid of energy of light. Thus, CO_2 is of primary importance for photosynthesis.

Method: Firstly the plant is destarched by keeping it in dark for about 48 hours. In a wide mouthed bottle filled 1/3 of its volume with the solution of caustic potash, half of a destarched green leaf is inserted and fixed inside the bottle with the help of split cork. Leaf should not touch caustic potash solution as shown in figure. The apparatus is kept in light for few hours. The leaf is tested for starch with iodine after keeping it in boiling water and decolorizing with alcohol.

Observations: The area of leaf outside of the bottle turns blue because of presence of starch in this region. The region of the leaf which was inside the bottle does not show any color change because of absence of starch in those regions. The outer portion of leaf shows presence of starch because it gets light and carbon dioxide and photosynthesis takes place.

Results: In the inner portion starch is not formed (from iodine or starch test) due to absence of carbon dioxide which is orbed by KOH thereby checking the photosynthesis. Considering above facts it becomes evident that carbon dioxide and light, both are essential for photosynthesis.

Precautions

1. Plants must be destarched.
2. Make the bottle air tight to check entry of CO_2 inside the bottle.

Exercise 19

To Demonstrate that O_2 is Evolved During Photosynthesis

Object: To demonstrate that O_2 is evolved during photosynthesis.

Material required: Hydrilla plant, beaker, test tube, funnel, pond water and pyrogallol.

Theory: Oxygen is a byproduct of photosynthesis, which is evolved in presence of light and chlorophyll during photosynthesis.

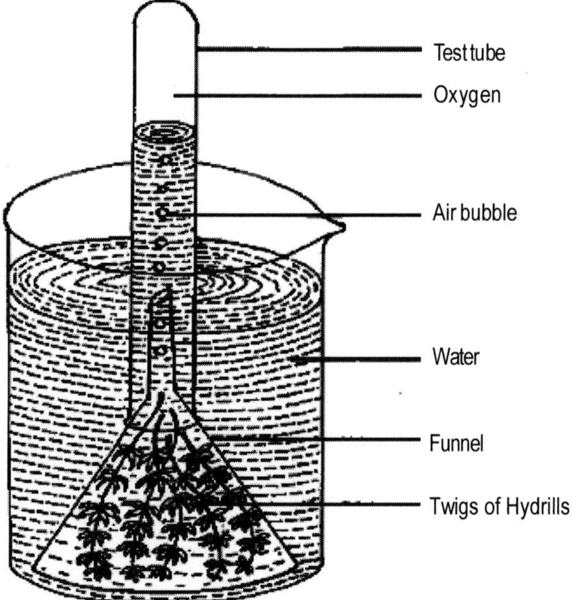

Fig. 14. Demonstration of O_2 evolved during photosynthesis.

$$6CO_2 + 12\ H_2O \xrightarrow[\text{Chlorophyll}]{\text{Light}} C_6H_{12}O_6 + 6H_2O + 6O_2$$

Method: Few twigs of Hydrilla plant are kept under inverter funnel in a beaker filled with pond water. The cut ends of the plant should face the tube of funnel. A test tube filled with water is placed upside down over the tube of the funnel partially dipped in the water of the beaker as shown in figure. This experimental set-up is kept under light for an hour.

Observations: After some time the air bubbles begin emerging (O_2) from the cut ends of the plant and collected at the top of the test tube with decrease in the volume of water in the test tube.

Test: A pyrogallol piece is introduced into the test tube which absorbs the O_2 immediately and the tube gets refilled with water. This indicates that gas getting collected is O_2.

Results: The decrease in volume of water in the test tube shows evolution of O_2 during photosynthesis, which accumulates in the test tube.

Precautions

1. Keep the end of the funnel under water.
2. Place the apparatus in light.
3. Remove the test tube carefully.
4. The cut end of the shoot should be projected upward.

Exercise 20

To Demonstrate Anaerobic Respiration

Object: To demonstrate anaerobic respiration.

Material required: Soaked pea or gram seeds, mercury, test tube stand, forceps and KOH pellets.

Theory: Generally, plants at certain stage have the capacity to respire in absence of oxygen i.e. anaerobic respiration. In this process, oxidation of carbohydrate takes places, ethyl alcohol and CO_2 being end products and only a little amount of energy is liberated-

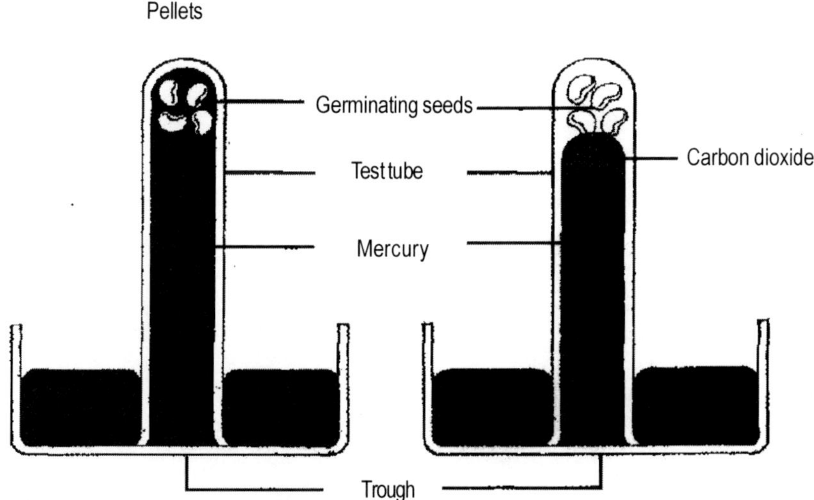

Fig. 15. Carbon dioxide evolution in anaerobic respiration

$C_6H_{12}O_6 - 2C_2H_5OH + 2CO_2 + 56Kcal$

Method: Soak some gram seeds in water. Remove their seed coat. A test tube completely filled with mercury is inverted in a petridish filled with mercury. Support the test tube with clamp. Now with the help of forceps, the seeds are introduced in the test tube through open end. The seeds being lighter in weight will rise to the top of the test tube. There is no air as shown in figure. This experiment is left as such for some time.

Observations: After some time the level of mercury in test tube falls due to the pressure exerted by the gas evolved from the seeds. With the help of a bent tube pass a little amount of caustic potash pellets, so that it floats in surface of mercury and absorb the evolved gas. The mercury level soon rises up again in the tube.

Results: Since the level of mercury falls down due to evolution of gas by the seeds, and on introducing the KOH pellets they absorb proving that it is CO_2. Hence, CO_2 is given out during anaerobic respiration.

Precautions

1. While introducing the seeds the entry of air bubble should be avoided.
2. The seed should be de-coated
3. In the beginning the test tube should be completely filled with mercury.

Exercise 21

To Demonstrate that Oxygen is Used up During Respiration

Object: To demonstrate that oxygen is used up during respiration.

Material required: A conical flask, some germinating seeds, glass tube having two bends at right angles, mercury, beaker, small tube and caustic potash (KOH).

Theory: Plants at some stage have a capacity to respire more rapidly in presence of oxygen i.e. aerobic respiration. In this process, oxidation of carbohydrate occurs resulting in conversion of carbohydrate into carbon dioxide (CO_2) and water with release of large amount of energy.

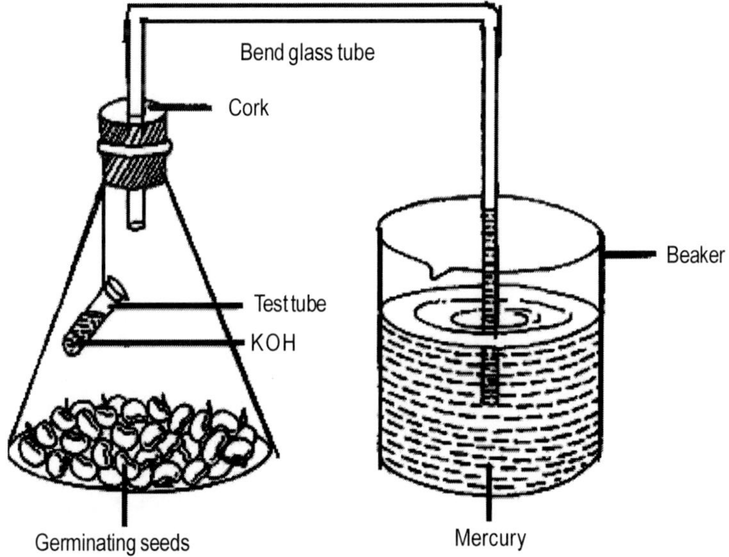

Fig. 16. Experiment to demonstrate aerobic respiration

$C_6H_{12}O_6$ $+6O_2$ -------------- $\leftarrow$ $6CO_2 + 6H_2O$ + energy (686 in K cal)

Method: Germinate some gram seeds in water and remove their seed coat. These germinating seeds along with KOH in a small tube are kept in a conical flask connected to the beaker filled with mercury through the glass tube with the help of cork as shown in figure. The apparatus is set air tight and the level of mercury is recorded before and after the experiment.

Observations

The level of the mercury rises in the glass tube.

Result: The germinating seeds respire, absorb oxygen and evolve carbon dioxide, and the evolved carbon dioxide is absorbed by caustic potash due to which volume of the air decreases and as a result the mercury level rises. It makes evident that O_2 is absorbed during respiration.

Precautions

(a) Use germinating seeds.

(b) Whole experiment should be air tight.

(c) Insert KOH inside the conical flask very carefully.

Exercise 22

To Determine Respiratory Quotient of Different Types of Respiratory Substrates

Object: To determine respiratory quotient of different types of respiratory substrates.

Material required: Ganong's respirometer, germinating seeds and saline water or mercury.

Theory: The respiratory quotient (R.Q.) is the ratio of CO_2 evolved to O_2 consumed i.e.

CO_2 / O_2.

$$R.Q. = \frac{\text{Volume of } CO_2 \text{ evolved}}{\text{Volume of } O_2 \text{ absorbed}}$$

Method: Ganong's respirometer is set on a stand. Saline water or mercury is filled from the side of leveling tube. Along with some moisture one type of respiratory substrate is placed in the respiratory chamber of the respirometer. The level in the graduated tube is made parallel to that of leveling- tube. The levels can be

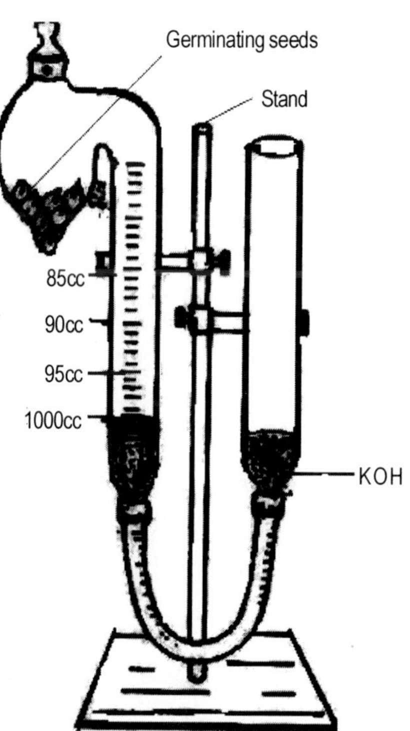

Fig. 17. Ganong's respirometer

adjusted with the help of leveling tube. The apparatus is made air tight by twisting the stopper cork of the chamber. The experiment is kept for a fixed time and the changes in the level of Saline water or mercury is recorded. Now a little KOH is introduced into the solution. The KOH diffuses and absorbs CO_2 released due to which level again changes. These changes are again recorded. The whole process is repeated for different respiratory substrates and observations are recorded separately.

Observations

(a) **In case of starchy seeds**: the first reading shows no change in level of Saline water or mercury in graduated tube. It indicates that: volume of O_2 absorbed is equal to volume of CO_2 evolved during respiration. After the addition of KOH the second reading however shows a rise in level due to absorption of CO_2 present in respiratory chamber by KOH.

(b) **In case of proteinaceous seeds and fatty seeds**: the first reading shows some rise in level indicting the lesser volume of CO_2 is produced as compare to O_2 consumed. The second reading shows further rise in level due to absorption of CO_2 by KOH.

(c) **In case of organic acid substrates**: first reading shows fall in level indicating the amount of CO_2 evolved is more as compare to O_2 consumed. The second reading after introducing KOH shows a rise in level above the original level.

Result: If there is no change in mercury level it indicates that the volume of CO_2 given out is equal to the volume of oxygen utilized in respiration. In this case value of R.Q. will be 1 if the mercury level falls it will indicate that CO_2 given out is more than O_2 consumed. In this case R.Q. will be more than 1, and if mercury level rises up it will indicate CO_2 given is less than O_2 consumed during respiration. In this case value of R.Q. will be less than 1.

Precautions

1. The apparatus should be air tight.
2. The level of saline water in graduated tube leveling tube should be parallel in the beginning of experiment.
3. Add only small quantity of KOH.

Exercise 23

To Demonstrate the Optimum Conditions for Seed Germination

Object: To demonstrate the optimum conditions for seed germination.

Material required: Beaker, glass slide, bean seeds, thread and water.

Theory: Germination is a process by which the embryo in the seed becomes activated and begins to grow into a new seedling under favorable conditions. Food is stored in the seeds in dry conditions but the developing embryo cannot utilize this dry food. Food can be utilized in liquid form only and seeds can utilize only dissolved oxygen. Excess water stop germination because once all the dissolves oxygen is utilized by the seed, further germination is not possible as the life supporting oxygen is lacking. Suitable temperature is necessary because low temperature retards the embryo activity and high temperature destroys the delicate embryo tissue. Seed usually germinate between 0°C to 50°C temperature and the optimum temperature is lies between 25 to 30°C. During germination rapid cell division takes place. Energy is required for cell division. This energy is obtained from oxidation. The oxygen is required for oxidation is supplied by the air.

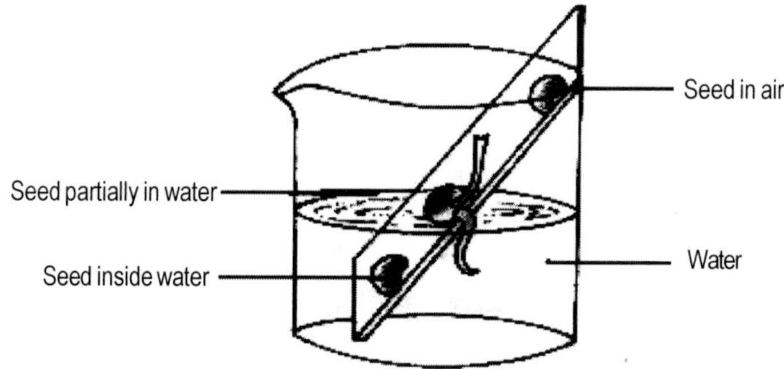

Fig. 18. Three bean seed experiment

Method: Three bean experiment; Tie three healthy and fresh bean seeds on a glass slide and keel this slide in water containing beaker in such a way that one seed is remain completely in water, the middle one is half immersed in water and upper seed is in air. Keep this set up at window, where sunlight is available .Observations are made after germination starts.

Observations: After few days we observe that germination has taken place in the middle water dipped seed.

Observation Table

S. No	Name of seed crop	Position of seed	Germination status
1.		Water dipped	
		Half water dipped	
		Above water level	
2.		Water dipped	
		Half water dipped	
		Above water level	

Results: After few days we observe that germination has taken place in the middle water dipped seed, as it gets all three optimum conditions i.e. water, temperature and air.

Precautions

1. Seeds should be fresh and healthy.
2. Seeds should be tight on glass slide loosely.
3. Seeds should be viable.

Exercise 24

Breaking of Seed Dormancy to Facilitate Germination

Object: Breaking of seed dormancy to facilitate germination.

Material required: Seeds, water, sharp needle, sharp razor, sandpaper, acids, Potassium nitrate and gibberellic acid.

Definition: Inability of a viable seed to respond the favorable environmental conditions for germination is known as dormancy.

Dormancy is mainly of following types

1. Hard seededness
2. Physiological dormancy
3. Presence of inhibitory substances.

Methods of breaking hard seededness

Hard seededness is the result of the impermeability of the seed coat to water or gases or physical resistance to embryo expansion whereas, inability of the seed to germinate because of unfavorable environmental conditions is known as quiescence.

Hard seediness in crops of family Leguminaceae and Malvaceae occur because of impermeability of water. No attempts are made to break this dormancy and dormant seeds are considered as germinated.

Dormant seeds are soaked in distilled water for 24-48 hrs and germination test is performed just after soaking. Example: rice.

Mechanical scarification: Seeds of some crops (fababean, rice bean etc.) are scarified mechanically to break the hard seediness by following methods

(i). **Piercing:** Seed coat is pierced with the help of a sharp needle without any injury to other parts of seed.

(ii) **Chipping**: Seed coat is chipped with a sharp razor without any injury to other parts of seed.

(iii) **Filing**: Seed coat is scratched at the suitable site with a file or sand-paper for entry of water.

Acid scarification: Seeds are soaked for prescribed period in concentrated sulphuric or nitric acid to make the seed coat pitted. After treatment seeds are thoroughly washed in running water and then tested for germination. Example: rice.

Methods for breaking physiological dormancy

Failure of viable mature embryo to germinate even when it is isolated from the seed coat is considered as physiological dormancy or embryo dormancy. This phenomenon is common in the woody members of family Rosaceae. It can be overcome by following methods.

Dry storage: Seeds of the crops having dormancy for short period of time are stored safely up to that period to break the dormancy Example: sunflower.

Stratification: Exposing, imbibed seeds to cool or warm temperature prior to germination in order to break the dormancy is known as stratification.

(i) Prechilling treatment: Seeds are arranged on a paper towel for germination test. These moistened paper towel containing seeds are kept at a low temperature (5-10^{0}C) up to 7 days before germinability test. Example: wheat, barley and onion.

(ii) Preheating treatment: Seeds are arranged on a paper towel for germination test. These rolled towels are heated by free air circulation with maximum 30-35^{0}C temperature for 7 days before placement for germinability test.

Example: Groundnut (40^{0}C), wheat, coat, barley, sunflower and berseem (30-35^{0}C) and rice (50^{0}C).

Light: Light with 750-1250 Lux intensity from Cool white lamps is illuminated for 8 hrs in every 24 hrs during the high temperature period to break the dormancy of grass seeds with alternate temperature. Example: Jussiae, Paulonia tomentosa and Kalanchoe blossfeldiana.

Potassium nitrate: A solution of KNO_3 (0.2%) is prepared by dissolving 2 g KNO_3 in 1 liter of water. The substratum selected for germination test is moistened with this solution instead of water. Seeds are arranged on it and kept for germination.

Example: Brassica, tomato, cole crops and chili.

Gibberelic acid: GA_3 solution of 0.05% is prepared by dissolving 500mg GA_3 in 1 liter water. Seeds are arranged on a substratum moistened with this solution and placed germinator.

Example: Oat, barley and wheat.

Sealed polythene envelope: The fresh non germinated seeds of berseem obtained after germination test are again arranged on a paper towel. It is kept in a sealed polythene envelope for the period and conditions prescribed for germination test to induce germinability.

Methods for removal of inhibitory substances

Some crops develop inhibitory substance on the surface of seed at the time of physiological maturity in field. These inhibitory substances have ability to check the germination of viable seeds. Removal of these substances by following methods from the seed results in normal germination.

Prewashing: Inhibitory substances from the surface of the seed coat are removed by washing the seed in running water at 25^{0}C. Seeds are then dried at room temperature prior to germination test.

Example: Beta, Xanthium and Fraxinus.

Removal of structure around the seed: Inhibitory structures like involucres of bristle lemma and palea, hard shell etc are removed with the help of forceps prior to germination test.

Exercise 25

Determination of the Seeds Viability by Tetrazolium Test

Object: Determination of the viability of seeds by tetrazolium test.

Material required: Healthy seeds, tetrazolium solutions, beakers, water and petridishes.

Theory: Viability of seeds shows their potential towards germination.

Method: Soak the seeds in water for overnight at room temperature. Then remove the seeds coat to facilitate the quick penetration of tetrazolium. After this, the seeds are kept in petridishes and soaked in 1% tetrazolium solution at pH 6-7 and kept in dark at about 30^0 C for 2-3 hrs. (for color development). As the color gets developed, drain the tetrazolium solution and rinse the seeds 2-3 times with water and evaluate the staining pattern. During evaluation, the seeds should be kept in water.

Observations: Red stained area indicates living tissues whereas unstained area represents dead tissue. If entire embryo is stained bright red it indicates viability of the seed. Count both types of seeds and calculate per cent viability by using the formula:

$$\text{\% Seed viability} = \frac{\text{No. of Viable seeds}}{\text{Total No. of seeds}} \times 100$$

Results: The development of red colour in the stained seeds gives the viability test.

Precautions

1. Seeds should be healthy prior to imbibitions.
2. Remove the seed coat very carefully without damaging the embryo.
3. Seeds should be thoroughly dipped in the chemical.

Exercise 26

To Demonstrate the Measurement of Growth in Plants with the Help of Arc Auxanometer

Object: To demonstrate the measurement of growth in plants with the help of Arc auxanometer.

Material required: Potted plant, Arc auxanometer, rope and weight.

Theory: Growth is a permanent change in plants with respect to its size, form, weight and volume.

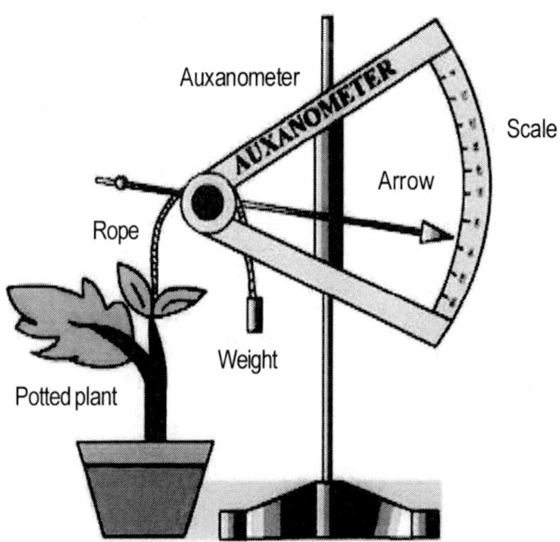

Fig. 19. An arc auxanometer

Method: An arc auxanometer consists of vertical stand with a pulley connected to a pointer on a graduated Arc as shown in figure. A thread is passed over the pulley with one end tied to the growing point of the plant and other end carrying a weight to keep the thread stretched. The whole experiment is kept for a day or two days and observed.

Observations: With the growth in plant, pulley moves down ward resulting in change in pointer showing movement on the graduated Arc scale.

Results: A downward movement of pointer on the graduated Arc scale indicates the rate of growth in the plants.

Precautions

1. Attach pulley with the growing point of the plant very carefully.
2. The plant should be potted and healthy.

Exercise 27

Growth Analysis: Calculation of Growth Parameters

Object: Growth analysis: Calculation of growth parameters.

Material required: Primary data collected from individual plants or derived from whole canopies.

Theory: The growth analysis study provides information on the growth of plants or of a plant community in a precise way with the available' raw data. The study also provides precise information on the nature of plant and environment interaction in a particular habitat.

Method

Components of growth analysis

1. Relative growth rate (RGR): This is the basic component of growth analysis which arose from the work of Blackman (1919). Blackman introduced the concept of compound interest law in growth.

RGR is defined at any instant of time (t) as the increase in dry weight per unit dry material present. This can be determined by measuring plant dry weight periodically during growth and is commonly represented as g g-1 day-1 or g g-1 week-1.

$$RGR = \frac{\text{Loge } W_2 - \text{Loge} - W_1}{t_2 - t_1}$$

Where W_1 and W_2 are the plant dry weight at times t_1 and t_2 respectively. **Drawback:** RGR does not give any knowledge about the size of assimilatory surface and the amount of photosynthetic that' is going into plant material. So in succeeding years, two other concepts were developed for use in growth analysis. These are NAR and LAR.

Actually RGR is the product of NAR and LAR

RGR = NAR x LAR

(2) Net assimilation rate (NAR)

Net assimilation rate is defined as the increase in plant dry weight per unit of assimilatory surface per unit time. NAR can be determined by measuring plant dry weight and leaf area periodically during growth and is commonly reported as g dm^{-2} $week^{-1}$.

$$NAR = \frac{W_2 - W_1}{t_2 - t_1} \qquad \frac{\text{Loge } LA_2 - \text{Loge } LA_1}{LA_2 - LA_1}$$

Where W_1 and W_2 are the plant dry weight at times t_1 and t_2 respectively.

LA_1 and LA_2 are the leaf area at times t_1 and t_2.

(3) Leaf area ratio (LAR)

LAR of a plant at any instant of time (t) is the ratio of total assimilatory surface to whole plant dry weight. This can be determined be measuring total leaf area and whole plant dry weight periodically and is commonly reported as cm^2 g-1

$$LAR = \frac{(LA_1/W_1) + (LA_2/W_2)}{t_2}$$

Where W_1 and W_2 are the print dry weight at times t_1 and t_2 respectively.

LA_1 and LA_2 are the leaf area at times t_1 and t_2.

LAR is the product of leaf weight ratio and specific leaf area.

LAR= LWR x SLA.

(4) Leaf weight ratio (LWR)

Leaf weight ratio is the ratio of dry weight of leaves to whole plant dry weight and is commonly reported as (g g-1).

$$L\,W\,R = \frac{WL}{PW}$$

Where WL = Dry weight of leaves.PW = Plant dry weight.

(5) Specific leaf area (SLA)

Specific leaf area is a measure of the leafiness of the plant on a dry weight basis and is commonly reported as (cm^2 g^{-1}). It is the ratio of leaf area to leaf dry weight.

$$SLA = \frac{\text{Leaf area}}{\text{Leaf dry weight}} \text{ expressed as g dm}^{-2}$$

(6) Specific leaf weight (SLW)

It is a measure of leaf weight per unit leaf area. Hence, it is a ratio.

$$SLW = \frac{\text{Leaf weight}}{\text{Leaf area}} \text{ expressed as g. dm}^{-2}$$

(7) Absolute growth rate (AGR)

AGR is the function of amount of growing material present and is influenced by the environment. It gives absolute values of biomass between two intervals. It is mainly used for a single plant or single plant organ e.g. leaf growth, plant height etc.

$$AGR = \frac{h_2 - h_1}{t_2 - t_1} \text{ cm day}^{-1}$$

Where h_1 and h_2 are the plant: height at times t_1 and t_2.

(8) Crop growth rate (CGR)

Crop growth rate represents total dry matter productivity of the community per unit land area over a certain time Span. Crop growth rate at any time can be determined by measuring the plant dry weight of a particular area at a regular interval of time divided by land area. It is commonly reported as kg m^{-2} $week^{-1}$ or kg m^{-2} $month^{-1}$.

$$CGR = \frac{W_2 - W_1}{(t_2 - t_1)} \times \frac{1}{L} \text{ g/m}^2\text{/day}$$

Where W_1 and W_2 are the plant dry weight at times t_1 and t_2.

L = Land area. Thus,

CGR = LAI x NAR.

(Actually LAI acts as a fulcrum between CGR and NAR).

(9) Leaf area index (LAI)

LAI may be defined as the area of leaves (one side only) divided by the ground area over which it is growing. Since leaf area and land area both have the same units, so LAI is expressed as a number.

$$LAI = \frac{LA}{L}$$

Where LA = Total leaf area L = Ground area

(10) Leaf area duration (LAD)

This is another component of growth analysis. This is the measure of the measure of the persistence of the assimilatory surface.

$$LAD = \frac{(A1 + A2)\ (t2 - t1)}{2}$$

Where A_1 and A_2 are leaf area index at times t_1 and t_2.

As this is the product of LAI and time, so the unit of LAD is days.

An approximate estimate of total yield (kg m^{-2}) can now be defined as the product of LAD x NAR. The estimate is approximate in the sense because in practice NAR cannot be determined accurately.

(11) Harvest index

It is the ratio of economic yield to the total biological yield expressed in percentage. It represents the efficiency of photosynthetic translocation to the economic parts.

$$\%\ \text{Harvest index} = \frac{\text{Economic yield}}{\text{Biological yield}} \times 100$$

Economic yield is the amount of productivity which is partitioned into the useful or harvested portion of the crop e.g. grains of cereals, the trunk of timber trees or the shoots of herbage crops.

Calculate RGR, NAR and LAR from the table given below

Observations: Total plant dry wt. and leaf area of a rice plant was recorded at 20 days interval" as follows.

Age in days	Plant dry wt.(g)	Leaf area(cm^2)
20	25.55	25.50
40	50.25	40.20
60	78.48	48.50
80	85.62	56.80
100	90.28	52.30

Result

(i) RGR = --

(ii) NAR = --

(iii) LAR = --

Exercise 28

Yield Estimation of Kharif Crops

Object: Yield estimation of kharif crops.

Material required: Polythene bags, balance, gunny bags or baskets, pans and harvesting equipment like sickle, knife etc.

Theory: Yield is the ultimate objective of cultivating agricultural plants. Different plant parts constitute the economic yield in different crops, like grain in cereals, pulses and oil seeds, vegetative parts or fruits in record yield at the correct maturation stage. Do not allow excessive wilting or drying of the material. Estimate yield in a sufficiently large sample. Materials required only a representative crop to estimate yields. (Do not select plants of any particular type. Do not damage any plant part vegetable crops; bark or out growth in fiber crops (jute, cotton); and the entire plant in fodder crops. The yield per unit area (per M^2 or ha) is the product of yield per plant multiplied by the number of plants per unit area. It is, therefore, essential to know the yield components per plant as well as per unit area.

Procedure

Single plant yield

Harvest 50 to 100 random single plants separately.

Count the number of branches or tillers, if any, for each plant.

Thresh the seed or economic produce separately from each branch or tiller or from the entire plant, as the case may be. Weigh the seed from each branch or tiller in grain crops and economic produce in vegetables, fodder and fiber crops and record it.

Determine average number of branches/tillers per plant, yield per branch/tiller.

Yield per unit area

Mark out 10 plots of 1 M^2 randomly in the field.

Count the total number of plants per plot. Harvest each plot separately.

Thresh the material at the appropriate stage. Record yield from each plot separately.

Observations

Single plant yield

Number of productive branches or tillers per plant.

Average number of branches or tillers per plant.

Weight of seeds per branch or tiller.

Average weight of seeds for each branch or tiller in g.

Average weight of seeds per plant in g.

Weight of 1000 seeds in small grains and 100 seeds in big grains in g.

Yield per unit area

Average number of plants per plot

Yield per plot of 1.0 sq. m (g)

Average yield per plot of 1 sq. m (g).

Calculations

Calculate the yield per hectare through the single plant yield method and yield per unit area method.

Yield per ha = Wt. of seeds per plant x No. of plants per ha. Or

Yield per plot of 1.0 sq. m x 10000

Problem 1: Calculate yield/ha of allotted crop adopting the above procedures.

Precautions

1. Use only a representative crop to estimate yields.
2. Do not select plants of any particular type.
3. Do not damage any plant part.

4. Record yield at the correct maturation stage.
5. Do not allow excessive wilting or drying of the material.
6. Estimate yield in a sufficiently large sample.

Exercise 29

Effect of Ethylene on Regulation of Stomata

Object: Effect of ethylene on regulation of stomata.

Material required: Plant leaves, ethaphon, ACC, double distilled water, measuring cylinder, beaker and electric balance.

Theory: Ethylene is a plant hormone that regulates many aspects of growth and development. Despite the well-known association between ethylene and stress signalling, its effects on stomatal movements are largely unexplored. Here, genetic and physiological data are provided that position ethylene into the Arabidopsis guard cell signalling network, and demonstrate a functional link between ethylene and hydrogen peroxide (H_2O_2). In wild-type leaves, ethylene induces stomatal closure that is dependent on H_2O_2 production in guard cells, generated by the nicotinamide adenine dinucleotide phosphate hydrogen (NADPH) oxidase AtrbohF.

Method: Ethylene induces stomatal closure in Arabidopsis; To determine the effect of ethylene on Arabidopsis stomata, wild-type leaves were floated on solutions containing various doses of ethephon(0.0,50.0 and 100.0 μm) or in ACC(0.0,50.0 and 100.0 μm) .The immediate precursor of ethylene. Stomatal apertures were then measured after 2.5 h. in epidermal leaf fragments prepared from these leaves. Both ethaphon and ACC (1-aminocyclopropane-1-carboxylic acid) caused stomatal closure in a dose-dependent manner. The ethephon-induced closure was initiated within 30 min of exposure. Furthermore, application of ethaphon to either soil- or hydroponically grown plants reduced transpiration.

Observation: It was observed that both ethephon and ACC (1-aminocyclopropane-1-carboxylic acid) caused stomatal closure in a dose-dependent manner.

Precautions

1. Leaves should be fresh.
2. The solution of ethaphon and ACC should be of proper concentration.
3. All glasswares and plasticwares should be neat and clean.

List of Colour Plates

Exercise 9: To Demonstrate Chlorophyll Fluorescence

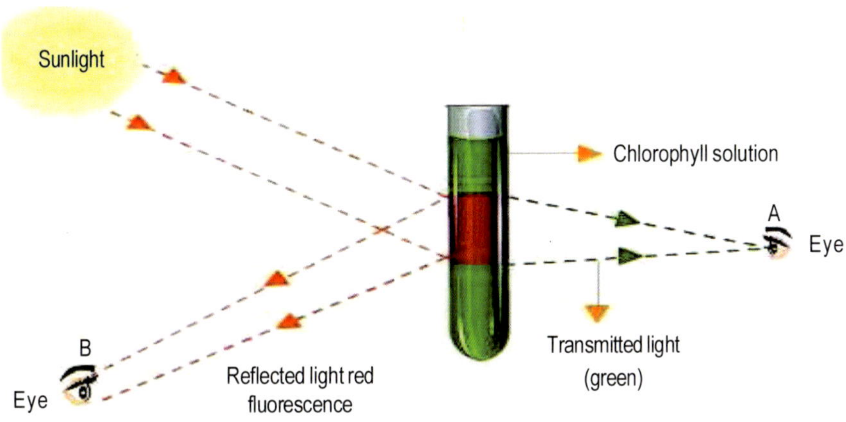

Fig. 6. Demonstration of chlorophyll fluorescence

Exercise 10: Measurement of Leaf Area by Graph Paper Method

Fig. 7. Graph paper

Exercise 12: Demonstration of Leaf Anatomy or Structure of C_3 and C_4 Plants (Demonstration by Already Prepared Slides)

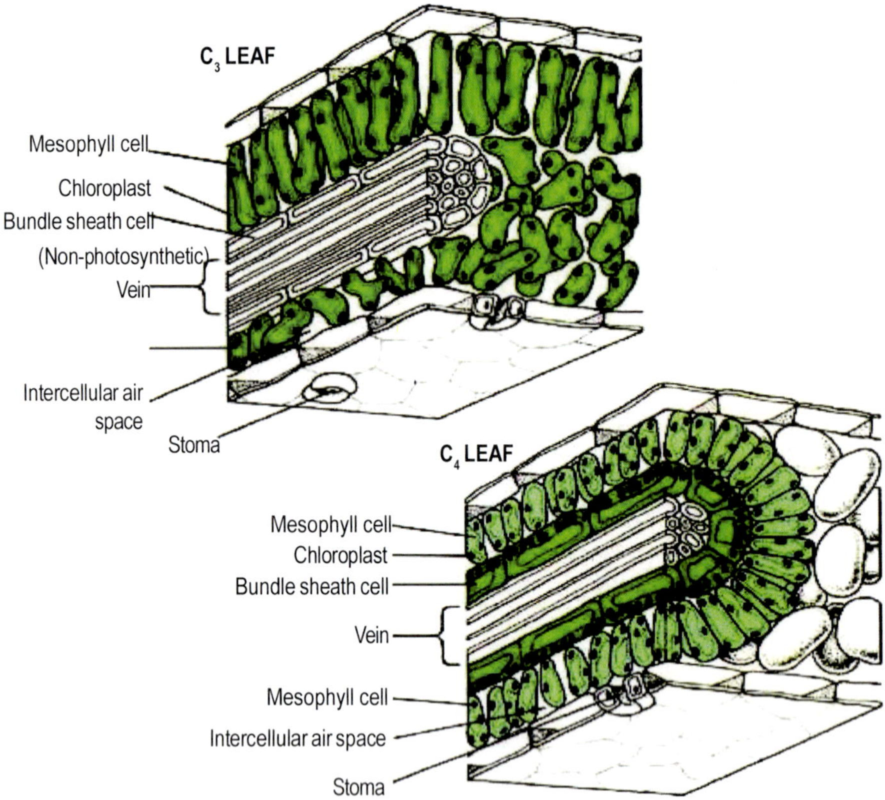

C3 LEAF
Mesophyll cell
Chloroplast
Bundle sheath cell
(Non-photosynthetic)
Vein
Intercellular air space
Stoma
C4 LEAF
Mesophyll cell
Chloroplast
Bundle sheath cell
Vein
Mesophyll cell
Intercellular air space
Stoma

Fig. 8. Leaf structure of C_3 and C_4 plants

Exercise 13: To Demonstrate that Water is Lost from a Living Plant During Transpiration

Water drops
Bell jar
Plant
Polythene cover
Vaseline
Glass plate

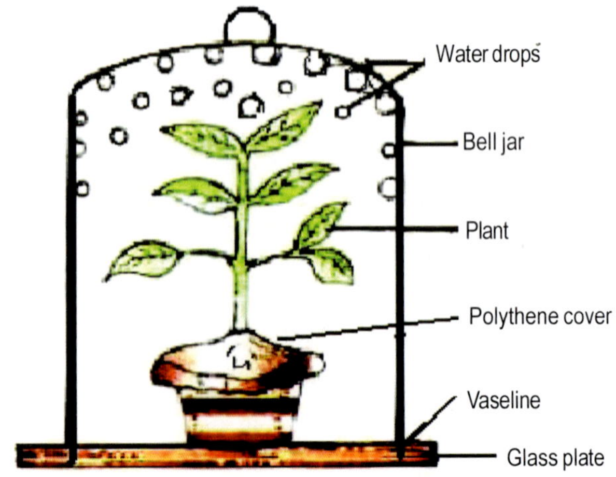

Fig. 9. Demostration of water loss by bell-jar

Exercise 15: To measure the Rate of Transpiration by Using Ganong's Potometer

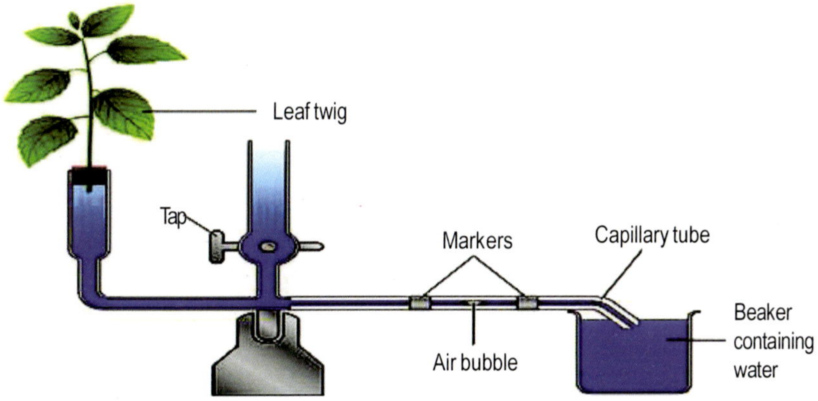

Fig. 10: Ganong's potometer

Exercise 26: To Demonstrate the Measurement of Growth in Plants with the Help of Arc Auxanometer

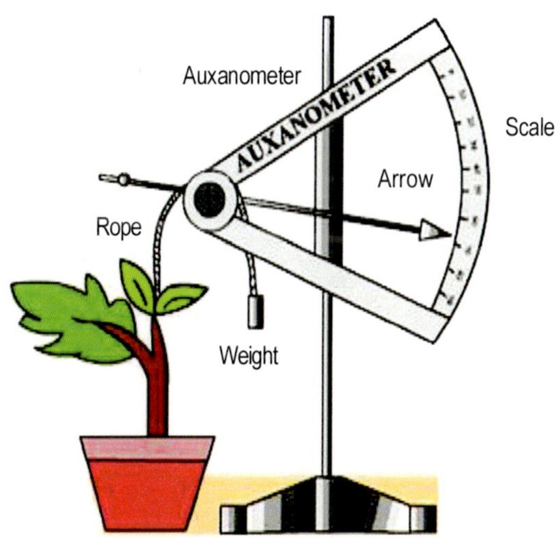

Fig. 19. An arc auxanometer

NIPA Publications on Plant Science

S.No.	Title	Author	ISBN	Year
1	A Colour Handbook on Practical Plant Pathology	Yadav, Vijay	9789385516177	2015
2	Abiotic Stress Tolerance in Crop Plants: Breeding and Biotechnology	Roy, Bidhan	9788189422943	2009
3	Advances in Soil Borne Plant Diseases	Naik, Manjunath	9788189422813	2008
4	Agricultural Biotechnology: Indian Print	Persley, G.J.: et al	9788189422097	2006
5	Agriculture Bioinformatics	Keshavachandran	9789383305421	2015
6	Agro-Informatics	Vanitha & Kalpana	9789380235714	2011
7	Air Pollution and It's Impacts on Plant Growth	Shyam, S. & H.N.Verma: et al	9788189422103	2006
8	An Introduction to Intellectual Property Rights	Pathak, Manju	9789383305124	2014
9	An Introduction to Nanotechnology	Rathinasamy, A.	9789381450413	2012
10	Applied Computational Biology and Statistics in Biotechnology and Bioinformatics (Set of 2 Vols.)	Roy, A.K.	9789380235929	2012
11	Approaches for Incorporating Drought and Salinity Resistance in Crop Plants	Chopra,V.L. & R.S.Paroda	9789383305742	2015
12	Auddaniki ke Adharbhoot Sidhant Tatha Nashijeev Prabandhan	Tiwari, A.K.	9789380235998	2012
13	Basics of Mutation Breeding	Thirugnanakumar,S.	9789383305193	2014
14	Biochemical Aspects of Plant Physiology	Bhattacharya, A. & Vijaya Luxmi	9789383305902	2016
15	Biochemistry,Molecular Biology and Biotechnology: Instant Notes	Gajera, H.P.,	9789383305520	2015
16	Biochemistry,Molecular Biology and Biotechnology: Instant Notes	Gajera, H.P.,	9789383305520	2015
17	Biochemistry,Molecular Biology and Biotechnology: Instant Notes	Gajera, H.P.,	9789383305520	2015
18	Bioinformatics in Agriculture: Tools and Applications	Balakrishnan,M.	9789381450925	2014
19	Biological Sciences: Innovations and Dynamics	Sinha,Rajeshwar P.	9789383305872	2015
20	Biotechnology in Horticulture: Methods and Applications	Peter, K.V.	9789381450918	2013
21	Biotechnology in India: Initiatives and Accomplishments	Niladri Bag	9789385516252	2016
22	Biotechnology of VA Mycorrhizza: Indian Scenario	Chandra, S. & H.K.Kehri	9788189422226	2006
23	Biotechnology: Practical Manual Series Vol 04	Thara, K.M.	9788190851237	2009
24	Breeding and Biotechnology of Flowers: Set of 2 Vols.	Singh, A.K.	9789383305612	2014
25	Breeding and Biotechnology of Flowers: Vol.1: Commercial Flowers	Singh, A.K.	9789383305353	2014
26	Breeding and Biotechnology of Flowers: Vol.2: Garden Flowers	Singh, A.K.	9789383305407	2014
27	Breeding and Protection of Vegetables	Rana, M.K.	9789380235493	2011
28	Breeding of Horticultural Crops: Principles and Practices: 2nd Revised & Expanded ed.	Kumar,N.	9789383305773	2015
29	Breeding,Biotechnology and Seed Production of Field Crops	Roy, Bidhan	9789381450680	2013
30	Climate Resilient Crops for the Future	Peter,K.V.	9789383305599	2015
31	Climatic Variability: Impacts on Agriculture and Allied Sectors	Datta, M.:Ed.	9789381450949	2014

S.No.	Title	Author	ISBN	Year
32	Computers in Agriculture: Fundamentals and Applications	Sharma Manish & Anil Bhatt	9789385516160	2015
33	Crop Diseases Management: Principles and Practices	Narayanasamy, P.	9789380235677	2011
34	Crop Diseases: Identification,Treatment and Management	Darwin, Henry	9789380235462	2011
35	Cyanobacteria: Antibacterial Activity	Kaushik, Purshotam	9788190723770	2009
36	Detritus and Decomposition in Ecosystems	Reshi, Zafar	9788189422158	2007
37	Developments in Physiology,Biochemistry and Molecular Biology of Plants Vol 01	Bose, Bandana & A.Hemantaranjan	9788189422028	2005
38	Developments in Physiology,Biochemistry and Molecular Biology of Plants Vol 02	Bose, Bandana & A.Hemantaranjan	9788189422929	2008
39	Diseases of Horticultural Crops Identification and Management: With Colour Illustrations	Kumar, Sanjeev	9789383305643	2015
40	Ecologically Based Integrated Pest Management (Set of 2 Vols.)	Abrol,D.P.& U.Shankar	9789380235950	2012
41	Elements of Entomology	H.L. Devasahayam	9789381450635	2014
42	Emerging Technologies of the 21st Century	Roy,A.K.	9789383305339	2015
43	Entomology: Novel Approaches	Jain,P.C. & M.C.Bhargava	9788189422325	2007
44	Experimental Biotechnology: Practical Manual Series 06	Dutta, Sunita & Abhijit Dutta	9789380235721	2011
45	Experimental Phytochemical Techniques	Raaman, N.	9789380235943	2012
46	Flower Crops: Cultivation and Management	Singh, A.K.	9788189422356	2006
47	Flowering Trees: Vol.12. Horticulture Science Series	Valsalakumari, P.K.	9788189422509	2008
48	Flowers for Trade: Vol.10. Horticulture Science Series	Sheela, V.L.	9788189422516	2008
49	Fruit Breeding	Dinesh, M.R.	9789383305513	2015
50	Genomics and Genetic Engineering	Satya, Pratik	9788189422776	2007
51	Handbook of Genetics and Biotechnology: Revised and Expanded 2nd ed.	Tomar, R.S.	9789383305445	2014
52	Heterosis Breeding in Vegetable Crops	Rai, Nagendra	9788189422035	2006
53	Identification and Management of Horticultural Pests	Ranjith, A.M.	9789381450567	2013
54	Illustrated Dictionary of Microbiology	Patel, Sunil	9788189422950	2008
55	Illustrated Plant Pathology: Basic Concepts	Darwin, Henry	9789380235080	2009
56	Improving Productivity of Drylands by Sustainable Resource Utilization and Management	Dayal, Devi	9789385516191	2016
57	Information Technology,Plant Pathology and Biodiversity: Indian Print	Bridge, P.D.	9788189422080	2005
58	Innovative Horticulture	Arunkumar, K.	9788189422738	2008
59	Integrated Pest Management in the Tropics (In 2 Parts)	Abrol, D.P.& U.Shankar	9789385516115	2016
60	Intellectual Property Rights Demystified	Ramkumar, Mu.	9788189422875	2008
61	Introductory Microbiology	Balachandar, D. & R.T.Vendan	9788189422783	2007
62	IPR: Drafting,Interpretation of Patent Specifications and Claims	Rathore,N.S.: Ed.	9789381450819	2013
63	Laboratory Manual of Biochemistry: Methods and Techniques	Sengar, R.S.	9789383305025	2014
64	Laboratory Manual of Microbiology: Practical Manual Series: 05	Roy, A.K.	9789380235189	2010
65	Merging Plant Breeding with Crop Biotechnology	Yasin, J.K.:Ed.	9789381450598	2012
66	Methods and Techniques in Plant Physiology	Bhattacharya, A. & Vijaya Luxmi	9789383305506	2015
67	Microbes: A Source of Energy for 21st Century	Soni, S.K.	9788189422141	2007
68	Microbial Biotechnology	Saikia, Ratul	9788189422806	2008

S.No.	Title	Author	ISBN	Year
69	Microbial Biotechnology for Sustainable Agriculture, Horticulture and Forestry	Bagyaraj, D.J.	9789380235820	2011
70	Microbial Diversity and Functions	Bagyaraj, D.J. & Tilak, K. et.al.	9789381450109	2012
71	Microbial Diversity and Its Applications	Barbudde, S.B.et al.	9789381450666	2013
72	Modern Biotechnology and Its Applications (Set of 2Vols.)	Behera, K.K.	9789381450833	2013
73	Modern Methods in Plant Physiology	Srivastava,G.Chand	9789380235011	2010
74	Molecular Biology and Biotechnology: Microbial Methods	Parakhia, M.V.	9789380235196	2010
75	Molecular Markers and Plant Biotechnology	Tomar, R.S.	9789380235257	2010
76	Nanotechnology in Agriculture	Subramanian, K.S.	9789383305209	2015
77	Nanotechnology in Soil Science and Plant Nutrition	Adhikari, Tapan	9789381450789	2013
78	Nematology: Fundamentals and Applications	Jonathan, E.I.	9789380235141	2010
79	Nitrogen Use Efficiency in Plants	Jain, Vanitha & P. Ananda Kumar	9789380235738	2011
80	Nutritional Disorders in Fruit Crops: Diagnosis and Management	Prakash, M.	9789381450956	2013
81	Objective Biochemistry	Sengar,R.S.	9789381450376	2014
82	Objective Genetics	Kumar S. Thirugnana	9789383305605	2015
83	Objective Microbiology	Nandi, S.	9789381450017	2011
84	Oilseeds: Properties,Products,Processing and Procedures	Nagaraj, G.	9788190723756	2009
85	Pest Management and Residual Analysis in Horticultural Crops	Gulati Rachna & Beena Kumari	9789381450710	2013
86	Pesticides: Methods of Their Residues Estimation	Kumari, Beena & T.S.Kathpal	9789380235394	2010
87	Pests of Vegetables: Bionomics and Management	Laskar, Nripendra	9789385516016	2015
88	Physio-Biochemistry and Biotechnology of Vegetable Crops	Rana, M.K.	9789380235318	2011
89	Physiological Disorders of Fruit Crops	Savreet, Sandhu & Bikramjit Singh Gill	9789381450581	2013
90	Phytochemical Techniques	Raaman, N.	9788189422301	2006
91	Phyto-nematodes in Crops: Their Identification, Treatment and Management	Sharma, G.L.	9789380235837	2011
92	Pigeonpea Hybrids and Their Production	A. N. Tikle,	9789383305957	2016
93	Plant Biochemistry: Techniques and Procedures	Nagaraj, G.	9789383305940	2015
94	Plant Diseases: Identification and Management (With Illustrations)	Rai, J.P.	9789383305315	2014
95	Plant Nutrient Disorders: Diagnosis and Management	Sarkar, A.K. & P.Mahapatra	9789385516023	2015
96	Plant Pathogens and Principles of Plant Pathology	Singh, Sanjeev	9789385516078	2016
97	Plant Secondary Metabolities	Shukla, Y.M. & R.Bhatnagar	9788190851220	2009
98	Plant Toxonomy and Biosystematics: Classical and Modern Methods	Rana,T.S.	9789383305414	2014
99	Plant-Microbe Interactions	Ramasamy and Kumar	9789383305834	2015
100	Practical Manual of Entomology)	Devasahayam,H.L.	9789380235905	2011
101	Practical Plant Biotechnology and Genetics	Rani, Archana	9789383305995	2015
102	Propagation of Horticultural Crops: Vol.06. Horticulture Science Series	Rajan, S. & B.L.Markose	9788189422486	2007

S.No.	Title	Author	ISBN	Year
103	Propagation of Horticultural Plants: Arid and Semi-Arid Regions	Singh, R.S. & R.Bhargava	9789383305254	2014
104	Quantitative Genetics and Crop Breeding	Thirugnanakumar, S.	9789380235981	2012
105	R Statistics	Samuel, Duleep	9789385516146	2015
106	Radio Tracer Techniques for Agriculturists and Biologists	Golakiya, B.A.	9788189422974	2008
107	Recent Advances in Biopesticides	Johri, Jayendra: eds.	9789380235219	2010
108	Systematics of Fruit Crops	Sharma, Girish	9789380235066	2009
109	Turning Plants Into Medicines: Novel Approaches	T.Parimelazhagan	9789381450468	2013
110	Turning Plants Into Medicines: Novel Approaches	T.Parimelazhagan	9789381450468	2013
111	Vegetable Crops: Genetics Resources and Improvements	D.K.Singh & H.Choudhary	9789380235509	2012
112	Vegetable Crops: Vol.04. Horticulture Science Series	Gopalakrishnan	9788189422417	2007
113	Weed Science	Das, P.C.	9789383305261	2015